Interfacial Electroviscoelasticity and Electrophoresis

Interfacial Electroviscoelasticity and Electrophoresis

Jyh-Ping Hsu

National Taiwan University

Taipei, Taiwan

Aleksandar M. Spasic

Institute for Technology of Nuclear and
Other Mineral Raw Materials

Belgrade, Serbia

CRC Press

Taylor & Francis Group

Boca Raton   London   New York

CRC Press is an imprint of the
Taylor & Francis Group an **informa** business

CRC Press
Taylor & Francis Group
6000 Broken Sound Parkway NW, Suite 300
Boca Raton, FL 33487-2742

First issued in paperback 2017

© 2010 by Taylor and Francis Group, LLC
CRC Press is an imprint of Taylor & Francis Group, an Informa business

No claim to original U.S. Government works

ISBN 13: 978-1-138-11390-9 (pbk)
ISBN 13: 978-1-4398-0352-3 (hbk)

Library of Congress Cataloging-in-Publication Data

Hsu, Jyh-Ping, 1955-
    Interfacial electroviscoelasticity and electrophoresis / Jyh-Ping Hsu, Aleksandar M. Spasic.
        p. cm.
    Includes bibliographical references and indexes.
    ISBN 978-1-4398-0352-3 (hardcover : alk. paper)
    1. Liquid-liquid interfaces. 2. Momentum transfer. 3. Electrophoresis. 4. Viscoelasticity. 5. Emulsions. I. Spasic, Aleksandar M. II. Title.

QD509.L54H78 2010
660'.2972--dc22   2009049466

Visit the Taylor & Francis Web site at
http://www.taylorandfrancis.com

and the CRC Press Web site at
http://www.crcpress.com

Contents

Preface

This book presents substantial insight into the development of the theory of electroviscoelasticity and electrophoresis and the motivation for developing this theory. Twenty years ago the idea, which was at first very foggy, developed from pilot plant experiments related to the extraction of uranium from wet phosphoric acid. In particular, the solution to the entrainment problems, the breaking of emulsions and double emulsions, as the succession after the extraction and stripping operations and processes, was performed. In this pilot plant, a secondary liquid-liquid phase separation loop was designed and carried out. The loop consisted of a lamellar coalescer and four flotation cells in series. Central equipment in the loop, relevant to this investigation, was the lamellar coalescer. The phase separation in this equipment is based on the action of external forces of mechanical and/or electrical origin, while adhesive processes at the inclined filling plates occur. Since many of the related processes (e.g., adhesive processes, rupture processes, and coalescence) were not very well understood, deeper research of these events and phenomena was a real scientific challenge.

This comprehensive reference describes recent developments in the basic and applied science and engineering of finely dispersed particles and related systems in general, but more profound and in-depth treatises are related to the liquid-liquid finely dispersed systems (i.e., emulsions and double emulsions). The book takes an interdisciplinary approach to the elucidation of the momentum transfer phenomenon as well as the electron transfer phenomenon, at well-characterized liquid-liquid interfaces. The considered scales are *micro*, *nano*, and *atto*; the micro scale may cover more or less classical chemical engineering insight, while the nano and atto scales focus on modern molecular and atomic engineering. In this context, *atomic engineering* recalls the ancient idea of interplay of particles that are small, indivisible, and integer (Greek $\alpha\tau o\mu o\zeta$). In the recent scientific literature, terms such as *nanoscience* and *nanotechnology*, *functional artificial nano-architectures*, *nanosystems*, and *molecular machinery*, once considered merely futuristic, have become foci of attention.

The aim of the proposed book is to provide the readers with recent concepts in the physics and chemistry of well-studied interfaces of rigid and deformable particles in homo- and heteroaggregate dispersed systems. Since many such systems are non-Newtonian, apart from classical momentum, heat, and mass transfer phenomena, the electron transfer phenomenon is also introduced into their description. Examples of such systems are emulsions, dispersoids, suspensions, nanopowders, foams, fluosols, polymer membranes, biocolloids, and plasmas. Thus, the central themes of this book are the hydrodynamic, electrodynamic, and thermodynamic instabilities that occur at interfaces; the rheological properties of the interfacial layers responsible for the existence of droplets, particles, and droplet-particle-film structures in finely dispersed systems; and the basic theory of electrokinetics, using the electrophoresis of rigid particles as an example.

The book is divided into six chapters. Chapter 1, "Introduction," contains classifications of finely dispersed systems based on various phenomenological notions: scales, geometry, the origin of forces, physical-chemical processes, and entities.

Chapter 2, "General," provides a historical review and motivation describing the pilot plant for uranium extraction from wet phosphoric acid. Thereafter, entrainment problems in solvent extraction underlined the performance of demulsions. There are also subsections related to the Marangoni instabilities and possible electrical analogies, and various constitutive models of liquids. This chapter ends by introducing the terms *electroviscosity* and *electroviscoelasticity* of liquid-liquid interfaces.

Chapter 3 focuses on the theory of electroviscoelasticity, including subsections: previous work; structure and electrified interfaces—a new constitutive model of liquids; and dynamics and physical formalism. Furthermore, three mathematical formalisms have been discussed: the stretching tensor model, van der Pol integer order derivative model, and van der Pol fractional order derivative model.

Chapter 4 emphasizes the experimental confirmation of previously postulated theoretical predictions in subsections describing the system, discussing the generation of the physical model, measuring the changes of the electrical interfacial potential at the liquid-liquid interface, and measuring resonant and/or characteristic frequencies of the representative system. Also included are results and discussion of the subsections, and assembled measured, calculated, and estimated data.

Chapter 5 discusses the implications and applications related to the first and second philosophical breakpoints. Finally, some conclusive remarks are given as well as a discussion of future potentials.

Because electrophoresis, one of the most important electrophoretic phenomena, is widely adopted either as an analytical tool to characterize the surface properties of colloid-sized particles or as a separation and purification process of both laboratory and industrial scales, the last chapter of this book focuses on the introduction of this phenomenon.

The intended audience of this book includes scientists, engineers, and graduate students with a basic knowledge of physical chemistry, electromagnetism, fluid mechanics, quantum mechanics, and wave mechanics.

Applications and implications of the material presented in the book are supposed to contribute to the advanced fundamental, applied, and engineering research of interfacial and colloidal phenomena. Broad subject examples include the following: (1) the first philosophical breakpoint may be related, for example, to the entrainment problems in solvent extraction operations and processes, colloid and interface science, chemical and biological sensors, electroanalytical methods, and/or biology and biomedicine (hematology, genetics, electroneurophysiology); (2) the second philosophical breakpoint may be used in deeper elucidation and research of, for example, spintronics, ionics, fractional-quantum Hall effect fluids, decoherence sensitivity, quantum computation, and entities and/or quantum particles entanglement.

Acknowledgments

This work was partially financially supported by the Ministry of Science and Environmental Protection of the Republic of Serbia, recently as a former Fundamental Research Project, "Multiphase Dispersed Systems," Grant no. 101822; at present as its continuation project, "Finely Dispersed Systems: Micro-, Nano-, and Atto-Engineering," Grant no. 142034; and in the past as a number of R&D projects.

Thanks are due to Professor A. S. Tolic, and to our colleagues M. Babic, M. Marinko, and N. Djokovic, all from the Institute for Technology of Nuclear and Other Mineral Raw Materials, Belgrade, Serbia, for help during the pilot plant development 22 years ago; the fundamental research presented here was initiated in this pilot plant. The efforts made by Dr. Li-Hsien Yeh in the preparation of, and consultation regarding, the electrophoresis part of this work are deeply appreciated.

Fruitful consultations, comments, discussions, and suggestions have been held with and received from Professor J. Jaric, Faculty of Mathematics, University of Belgrade, Serbia; and Professors M. Plavsic, M. V. Mitrovic, and D. N. Krstic, Faculty of Technology and Metallurgy, Belgrade, Serbia.

Many published papers and presentations related to the application of fractional calculus were realized in collaboration with Professor Mihailo P. Lazarevic, Faculty of Mechanical Engineering, University of Belgrade, Serbia.

Substantial encouragement, help, and support during the last two decades have been received from the late Professor Jaroslav Prochazka, Institute of Chemical Process Fundamentals, Czech Academy of Science, Prague, Czech Republic.

Important reviews of the edited book draft of this manuscript, and the related general idea, came from Professors H-J. Bart, Universität Kaiserslautern, Lehrstuhl für Thermische Verfahrenstechnik, Kaiserslautern, Germany; A. V. Delgado, Department of Applied Physics, Faculty of Science, University of Granada, Granada, Spain; H. Ohshima, Faculty of Pharmaceutical Sciences, Science University of Tokyo, Chiba, Japan; and A. T. Hubbard, Santa Barbara Science Project, California, United States.

And, last but not least, invited courses, lectures, and deeper collaboration arrived from Professors S. Tseng, Department of Mathematics, Tamkang University, Taipei, Taiwan; S. Alexandrova, Laboratoire de Thermique, Energétique et Procédés, Ecole Nationale Supérieure en Génie des Technologies Industrielles, Pau, France; and A. Saboni, Laboratoire des Génie des Procédés et Environnement, INSA de Rouen, Mont-Saint-Aignan, France.

Symbols

a	area (m^2)
A	amplitude (m)
c	velocity of light (m/s)
C	capacitance (Coul/V)
C_m	rate constant (A^2/m/s)
d	diameter (m)
e	electric charge (Coul)
E	electric field strength (V/m)
f	frequency (1/s)
f_i	force (N)
g	gravitational acceleration (m/s^2)
G	spring constant (kg/m/s^2)
ΔG	free energy (J)
h	Planck constant (Js)
i	electric current (A)
j	number of the identical oscillators (clusters) (–)
k	Boltzmann constant (J/K)
L	inductance (VsA)
m	mass (kg)
n	number (–)
N	number of elements (–)
p	pressure (N/m^2)
r	radius (m)
r_c	continuum resistance constant (kgs)
R	resistance (V/A)
Q_m	factor of merit (–)
Q_N	configuration integral (–)
$S_{i,e}$	internal and external separation, respectively (m)
ΔS	entropy (J/K)
t	time (s)
T	absolute temperature (K)
u	velocity (m/s)

U	electric potential (V)
V	volume (m^3)
w	energy density (J/m^3)
W	energy of the mechanical wave (J/m^3)
W_N	Boltzmann probability factor (–)
Z	impedance (V/A)
Z_p	partition function (–)

Greek Letters

α	linear contribution constant (A/V)
α^0	surface (–)
β	ratio (–)
β_T	uniform temperature gradient (K/m)
γ	nonlinear contribution constant (A/V^3)
γ_T	uniform potential gradient (V/m)
γ	(italic) displacement of the spring (m) (only in Section 2.1.4)
ε	dielectric constant (–)
λ	free path between two collisions (m)
λ	relaxation, retardation time (s)
ρ	density (kg/m^3)
μ	viscosity (kg/m/s)
μ_e	magnetic permeability constant (N/A^2)
μ_0	relative magnetic permeability (–)
κ^0	surface (–)
κ	heat diffusivity (m^2/s)
ν_v	kinematic viscosity (m^2/s)
τ	tangential tension (N/m^2)
σ_{ep}	electrical interfacial potential (J/Coul)
σ_{in}	interfacial tension (N/m)
σ	(italic) force (N)
ϕ	potential energy (J)
ϕ_e	electron flux density (Coul/m^2)
ω	angular frequency (1/s^2)
Ω	frequency of the incident oscillations (1/s)

Subscripts

a	approach
b	bulk
bo	buoyancy
c	coalescence
col	collapse
d	droplet
dist	disturbance
dom	dominant
e	electric
f	free
fu	"flow up"
g	gravitational
l	electromagnetic/Lorentz
o	resonant
R	rupture
RE	rest
s	surface

Superscripts

b	bulk
o	overall
s	surface

Abbreviations

A_n	spectral distribution of the noise current
AD_R	electrical critical parameter of the first order
AD_E	electrical critical parameter of the second order
DTK	D2EHPA-TOPO-kerosene
dA	elemental surface
D/Dt	substantial derivative
E	elastic
EDL	electrical double layer
EIP	electrical interfacial potential

F	force/vector
H	heavy liquid phase
L	light liquid phase
M	concentration
Ma	Marangoni number of the first order
Ma''	Marangoni number of the second order
NMR	nuclear magnetic resonance
R	rigid
$\sim$	vector, tensor
$\alpha_{0,1,2,3}$	resistance functions
δ	Kronecker symbol

1

Introduction

1.1 Classification of Finely Dispersed Systems

Over the last decade, the biggest advances in physics, physical chemistry, and biochemistry have come from thinking smaller. This research takes an interdisciplinary approach to the elucidation of the momentum transfer phenomenon as well as the electron transfer phenomenon at well-characterized, developed, both rigid and deformable liquid-liquid interfaces. The considered scales are micro, nano, and atto, using various theoretical approaches. Micro scales may cover more or less classical chemical-engineering insight, while nano and atto scales focus on modern molecular and atomic engineering. This chapter presents a brief overview of the various possible classifications of finely dispersed systems based on scales, geometry, the origin of forces, and physical-chemical processes. At the end of this chapter, a new classification of finely dispersed systems based on entities will be presented [1].

1.1.1 Classification Based on Scales

1.1.1.1 Macro and Micro Scales

Classical chemical engineering has been intensively developed during the last century. Theoretical backgrounds of momentum, mass, energy balances, and equilibrium states are commonly used as well as chemical thermodynamics and kinetics. Physical and mathematical formalisms are related to heat, mass, and momentum transfer phenomena as well as homogeneous and heterogeneous catalysis. Entire object models, continuum models, and constrained continuum models are frequently used for the description of the events, and for equipment designing. Usual, principal equipment are reactors, tanks, and columns. Output is, generally, demonstrated as conventional products, precision products, chemistry (solutions), and biochemistry.

1.1.1.2 Nano Scale

Molecular engineering nowadays still suffers from a lack of substantial development. Besides heat, mass, and momentum transfer phenomena, commonly used in classical chemical engineering, it is necessary to introduce the electron transfer phenomenon. A description of the events is based on molecular mechanics, molecular orbits, and electrodynamics. Principal tools and equipment are micro reactors, membrane systems, microanalytical sensors, and microelectronic devices. Output is, generally, demonstrated as molecules, chemistry (solutions), and biochemistry.

1.1.1.3 Atto Scale

Atto engineering, which has already existed for more than a whole century, is in permanent and almost infinite development. Its theoretical background is related to surface physics and chemistry, quantum and wave mechanics, and quantum electrodynamics. Discrete models and constrained discrete models are convenient for descriptions of related events. Tools and equipment are nano and atto dispersions and beams (demons, ions, phonons, infons, photons, and electrons), ultrathin films and membranes, fullerenes and buckytubules, Langmuir-Blodgett systems, molecular machines, nanoelectronic devices, and various beam generators. Output is, generally, demonstrated as finely dispersed particles (e.g., plasma, fluosol-fog, fluosol-smoke, foam, emulsion, suspension, metal, vesicle, and dispersoid).

1.1.2 Classification Based on Geometry

1.1.2.1 Surface Continuums

The hypothetical continuous material may be called a *continuous medium* or *continuum*. The adjective *continuous* refers to the concept of when the molecular structure of matter is disregarded and is without gaps or empty spaces. Also, it is supposed that mathematical functions entering the theory are continuous functions, except possibly at a finite number of interior surfaces separating regions of continuity [2]. Boundaries of condensed homogeneous phases (liquids and some solids) in the nano scale also present continua, but only with two dimensions, while in the third dimension their characteristics change. Such boundaries may be defined as bonded surface continua. Besides a surface tension force, all characteristics of nano layers at boundaries of condensed phases are different than their bulk properties (e.g., density, viscosity, heat, and mass transfer rates) in the surface toward bulk direction. Further on, in the nano space outside a condensed phase, another layer with interacting forces appears. Molecules and ions inside this nano space are attracted or repulsed, and attracted molecules or ions behave either as two-dimensional gasses or as another condensed nano phase. In both cases, characteristics (e.g., polarity, reactivity, and stereo position) of these adsorbed molecules or ions are different than the properties of "free" molecules or ions, and may be defined as other two-dimensional nano continua.

1.1.2.2 Line Continuums

Generally, bonded two-dimensional (surface) continua exist at the contact of two immiscible homogeneous phases. At the contact of three immiscible homogeneous phases, only line nano elements exist. Examples include contact lines between two liquids and one gaseous phase. Line elements are defined as one-dimensional bonded nano continua, with line tension forces, molecules, or ions serving as attraction or repulsion forces in the nano space over a line element and a complex asymmetric structure. Gasses adsorbed at a line element have only one degree of freedom (one-dimensional gasses; also can be condensed matter). Molecules or ions at line elements also change their characteristics.

1.1.2.3 Point Discontinuums

Bonded point elements exist at contacts of four phases, analogous to line elements at contacts of two liquid, one gaseous, and one solid phase. These elements are discontinuous in all dimensions. In the nano space over point elements, adsorbed molecules or ions have zero degrees of freedom.

1.1.3 Classification Based on Origin of Forces

Depending on the scale, it is possible to choose several different approaches to the classification based on forces and their origin (e.g., intersurface forces, interparticle forces, intermolecular forces, interatomic forces, and finally interentity forces). Here, the convenient classification may be based on four principal groups of forces acting at small separations: these are classified as DLVO (van der Waals forces and electrostatic forces) and non-DLVO (solvation forces and steric/fluctuation forces); *DLVO* stands for Deryagin-Landau and Verwey-Overbeek. More details related to the forces and their interactions in various combinations and situations may be found, for example, in Israelachvili's book *Intermolecular and Surface Forces* [6] and/or in Drexler's book *Nanosystems: Molecular Machinery, Manufacturing, and Computation* [7].

1.1.3.1 Electrostatic Forces

Electrostatic repulsive forces appear between the objects that have a similar electrical charge. The origin of a charge (e.g., in emulsions) may be explained by ionization and adsorption mechanisms. Hence, a double electrical layer is formed when the counterions and coions diffuse toward the surface and in the near-surface positions. A change of potential energy of interactions between, for example, two droplets or particles at some distance (e.g., ca. 30–40 nm or more) is given by the DLVO theory. Further on, for smaller separations when atomic electron clouds overlap, a strong Born repulsion appears.

1.1.3.2 van der Waals Forces

Neutral molecules contain some attractive forces caused by electrical interactions between three types of dipole configurations. The attraction occurs from dipole orientation between two permanent dipoles, between dipole and induced dipole, and/or between induced dipole and induced dipole. For the case of induced dipole and induced dipole, these forces are called *London dispersion forces*. Energy of attraction may be expressed using a Hamaker constant, which depends on the density and polarizability of the atoms in the particles. Typical values of this constant are situated between 10^{-20} and 10^{-19} J.

1.1.3.3 Solvation Forces

Solvation interactions can operate over multiple molecular diameters: in water, the repulsion between hydrophilic surfaces [6,7] falls off exponentially, becoming acceptably small at separations of 1–3 nm (hydrophobic forces would cause strong adhesion between surfaces and must be avoided). Oscillating solvation forces associated with molecular-sized effects become small at separations greater than ca. 2 nm in water [6,7], and are reduced by surface roughness and mixed-solvent systems. With rough bead surfaces and separation of ca. 6 nm, oscillating forces should not be observed [7].

1.1.3.4 Steric and Fluctuation Forces

Short-range repulsive forces are caused mainly by entropic effects. This mechanism is termed *steric stabilization* and is widely used to prevent the coagulation of colloids. Attachment of these chains renders the bead surfaces hydrophilic and inhibits the bead-bead adhesion that would otherwise occur though short-range hydrophobic and van der Waals attractions. Since the interacting surfaces are not smooth and rigid, the interfaces may be considered thermally diffuse, rather dynamically rough than statically rough [6].

1.1.4 Classification Based on Physical-Chemical Processes

Basically, there are four kinds of physical or chemical processes in nano continua, discontinua, and spaces of interactions: (1) processes in the nano spaces over surface, line, or point elements; (2) processes in the nano spaces inside surface, line, and point elements; (3) carrier processes—mass and energy transfer processes, with or without physical and chemical transformations, through internal and external nano spaces; and (4) membrane processes—mass and energy transport processes through membranes (double gas, liquid and solid surface, and line and point elements). Also, it is possible to present

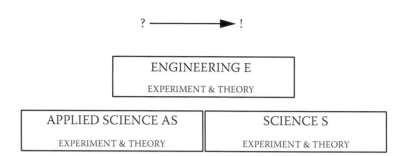

FIGURE 1.1
Research philosophy.

these processes in the following order: (1) diffusion, (2) sorption processes, (3) heterogeneous catalysis, and (4) membrane processes [8–10].

1.1.5 Classification Based on Entities

1.1.5.1 Research Philosophy

One brief answer to the questions "What?" and "Why?" (i.e., philosophical questions) could be schematically presented using the proposition "Along the path from a question mark toward an exclamation mark" (Figure 1.1).

> **E** (engineering): Takes the behavior as given and studies how to build the object that will behave as planned. Engineering enables things to work (e.g., the process, operation, and/or equipment *B* can fulfill the objective *C*); proof is the demonstration!
>
> **AS** (applied science): Theories related to the particular classes of things, just different from the science that studies general or universal laws. Some processes, operations, and/or equipment *B* can fulfill the objective *C*; when the demonstration becomes possible, then the applied science disappears!
>
> **S** (science): Takes the object as given and studies its behavior; science discovers and explains how things work!

1.1.5.2 Research Strategy and Methodology

Using the same proposition "Along the path question mark toward exclamation mark," one brief answer to the question "How?" (i.e., a strategic question) could be schematically presented following a methodological weight hierarchy (capability/feasibility), as is shown in Figure 1.2, although the researcher is not always aware of the strategic path or the step he or she is using at the time!

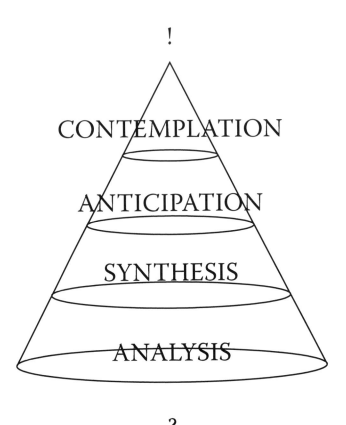

FIGURE 1.2
Research strategy and methodology.

1.1.5.3 Characteristics and Approximation and Abstraction Levels

A new idea, using the deterministic approach, has been applied for the eluci-dation of the electron and momentum transfer phenomena at both rigid and deformable interfaces in finely (micro, nano, and atto) dispersed systems. Since the events at the interfaces of finely dispersed systems have to be con-sidered at the molecular, atomic, and/or entity level, it is inevitable to intro-duce the electron transfer beside the classical heat, mass, and momentum transfer commonly used in chemical engineering [11]. Therefore, an entity can be defined as the smallest indivisible element of matter that is related to the particular transfer phenomena. Hence, the entity can be a differential element of mass/demon, an ion, a phonon as quantum of acoustic energy, an infon as quantum of information, a photon, or an electron [12].

A possible approach is proposed to the general formulation of the links between the basic characteristics, levels of approximation, and levels of abstraction related

TABLE 1.1

A New Classification of Finely Dispersed Systems

DP* DM*	Gas	Liquid	Solid
Gas	PLASMA D	FOAM D	METAL R
Liquid	FLUOSOL/fog D	EMULSION D	VESICLE D
Solid	FLUOSOL/smoke R	SUSPENSION R	DISPERSOID R

Note: DP* = Dispersed phase; DM* = dispersion medium.

to the existence of finely dispersed systems (DS) [12]. At first, for the reason of simpler and easier physical and mathematical modeling, it is convenient to introduce the terms *homoaggregate* (HOA; phases in the same state of aggregation) and *heteroaggregate* (HEA; phases in more than one state of aggregation). Now, the matrix presentation of all finely dispersed systems is given by

$$[(DS)^{ij} = (HOA)\delta^{ij} + (HEA)\tau^{ij}] \tag{1.1}$$

where i and j refer to the particular finely dispersed system position (i.e., when $i = j$, then diagonal positions correspond to the homoaggregate finely dispersed systems: plasmas, emulsions, and dispersoids, respectively), and when $i \neq j$, then tangential positions correspond to the heteroaggregate systems (fluosols/fog, fluosols/smoke, foam, suspension, metal, and vesicle, respectively). Furthermore, the interfaces may be deformable (*D*) and rigid (*R*), which is presented in Table 1.1.

Now, related to the levels of abstraction and approximation, it is possible to distinguish continuum models (the phases considered as a continuum, i.e., without discontinuities inside the entire phase, homogeneous, and isotropic) and discrete models (the phases considered according to the Born-Oppenheimer approximation: entities and nucleus/CTE motions are considered separately). Continuum models are convenient for microscale description (entire object models), for example, conventional products, precision products, chemistry/solutions, and biochemistry; while discrete models are convenient for nanoscale description (molecular mechanics, or molecular orbits), such as chemistry/solutions, biochemistry, and molecular engineering, and/or for atto-scale description (quantum electrodynamics), such as molecular engineering and atto engineering. Since the interfaces in finely dispersed systems are very developed, it is sensible to consider the discrete models approach for a description of related events [12]. For easier understanding, it is convenient to consult, among others, references [13–26].

1.1.5.4 Hierarchy of Entities

Figure 1.3a shows a stereographic projection/mapping from the Riemann sphere; Figure 1.3b shows a "hierarchy" of entities, which have to be understood as a limit value of the ratio u_0:Z (withdrawn from magnetic Reynolds

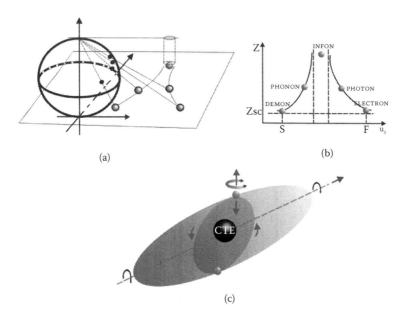

(a)

(b)

(c)

FIGURE 1.3

Entities.

Note: a: A stereographic projection/mapping from the Riemann sphere; b: a hierarchy of entities, vs. correlation viscosity/impedance = characteristic velocity, S = slow/demon (superfluid), and F = fast/electron (superconductor); and c: an entity as an energetic ellipsoid (at the same time macroscopic and microscopic), where CTE = center of total energy, motions (translation, rotation, vibration, precession, and angle rotation).

Source: From Spasic, A. M., Mitrovic, M., Krstic, D. N., Classification of finely dispersed systems, in: *Finely Dispersed Particles: Micro-, Nano, and Atto-Engineering.*, A. M. Spasic and J. P. Hsu, Eds., CRC Press/Taylor & Francis, Boca Raton, FL, 2006, p. 20, courtesy of CRC Press/Taylor & Francis.

criteria [$Re_m = 4\pi l G u_0/c^2$], where the conductivity G is expressed as a reciprocal of viscosity/impedance Z ($G = 1/Z$), 1 is the path length that an entity "overrides," u_0 is the characteristic velocity, and c is the velocity of light).

In general, S corresponds to the slow system or superfluid, and F corresponds to the fast system or superconductor; now, it is possible to propose that all real dynamic systems are situated between these limits. Also, it seems sensible to think about the further structure of entities; for example, the basic entity can be understood as an energetic ellipsoid, shown in Figure 1.3c (based on the model of electrons following Maxwell-Dirac isomorphism, or MDI: an electron is an entity at the same time that quantum-mechanical/microscopic $N = -2$ and electrodynamics/macroscopic $N = 3$) [5].

2

General

2.1 Historical Review and Motivation

This work was initiated in an attempt to apply electromechanical principles for the elucidation of the secondary liquid-liquid phase separation problems, methods, equipment, and/or plant conception. In solvent extraction operations during the generation of polydispersed systems, some kind of secondary liquid-liquid droplets (emulsions) or droplet-film structures (double emulsions) occur as an undesirable consequence. These droplets or droplet-film structures are small and stable, and, therefore, they are mechanically entrained by one of the primarily separated liquid phases. For the separation or breakage of simple or double emulsions, an additional force is needed. Figure 2.1 show a drop-size distribution and the critical diameters d_1 and d_2, which correspond to the boundaries of the primary and the secondary separation, respectively.

The primary separation is determined by the physical processes of gravitational sedimentation and coalescence. Subsequently during the secondary separation (of the, e.g., double emulsions), the processes of droplet-film rupture and coalescence occur in succession.

The aim of this work is to develop a general and flexible methodology which may become a purposeful tool in solving the secondary liquid-liquid phase separation problems. The work also intends to perform the secondary liquid-liquid phase separation unit on a pilot plant scale, and to compare its efficiency with theoretical predictions and pilot plant scale experience.

Successful prediction of a phase separation operation from any mathematical model requires determination of the necessary parameters of the system. These parameters can be determined through the use of theoretical correlations and/or experiments carried out on a laboratory or pilot plant scale. The reliability of determined parameters for use in the secondary liquid-liquid phase separation method selection and equipment design can be questionable, fair, and/or good. Good prediction of performance of phase separation units within a certain range of operating conditions requires, very often, the satisfaction of few contradicting conditions.

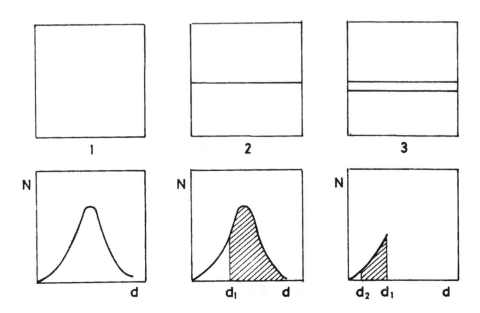

FIGURE 2.1
Drop-size distribution and critical diameters: (1) generation, (2) the primary separation, and (3) the secondary separation.

The particular problem considered as the representative one, while developing the knowledge and database, was the mechanical entrainment of one liquid phase by the other in the solvent extraction operation. Experimental results obtained in the pilot plant for uranium recovery from wet phosphoric acid were used as the comparable source [27–42]. In this pilot plant also, the secondary liquid-liquid phase separation loop has been realized [43–45]. The loop consisted of a lamellar coalescer and four flotation cells in series. Central equipment in the loop, relevant to this investigation, is the lamellar coalescer. The phase separation in this equipment is based on the acting forces of mechanical and/or electrical origin, while the adhesive processes at the inclined filling plates occur [46–51].

2.1.1 Pilot Plant for Uranium Extraction from Wet Phosphoric Acid

2.1.1.1 Description of the Physical-Chemical System

The light continuous liquid was a synergistic mixture of 0.5 M (di-(2 ethylhexil)) phosphoric acid—0.125 M trioctyl phosphin oxide in dearomatized kerosene (D2EHPA and TOPO manufactured by SNPE, France), and the heavy dispersed liquid was 5.6 M phosphoric acid. The structural presentation of the constituent liquids is as follows:

$$\begin{array}{ccc}
\mathrm{OH} & \mathrm{OH} & \mathrm{R'} \\
| & | & | \\
\mathrm{HO\text{-}P{=}O} & \mathrm{RO\text{-}P{=}O} & \mathrm{R'\text{-}P{=}O} \\
| & | & | \\
\mathrm{OH} & \mathrm{OR} & \mathrm{R'} \\
\end{array}$$

$$\begin{array}{ccc}
\mathrm{H_3PO_4} & \mathrm{D2EHPA} & \mathrm{TOPO}
\end{array}$$

where R is $CH_2\text{-}CH(CH_2)_3CH_3$, and R′ is $(CH_2)_7CH_3$.
$$|$$
$$C_2H_5$$

Figure 2.2 shows measured variations of the relevant physical properties of the liquids with temperature.

2.1.1.2 Droplet-Film Structure: Double Emulsion

The secondary separation of the entrained light phase or the breaking of emulsion was examined in the pilot plant [27, 28]. The schematic flow sheet of the selected operations, processes, and equipment is presented in Figure 2.3. In this pilot plant, the light-phase loop, relevant to this investigation [43,45], is shown in Figure 2.4.

The polydispersion was generated and the primary separation performed in a countercurrent "pump-mix" mixer-settler battery (numbered 2 in

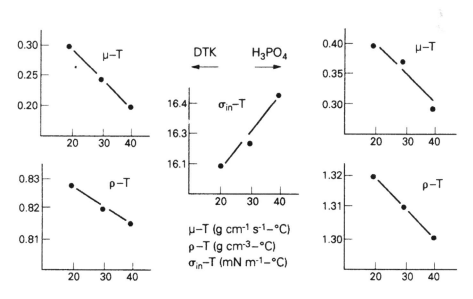

FIGURE 2.2
Measured variations of the relevant physical properties of the liquids with temperature, density, viscosity, and interfacial tension.

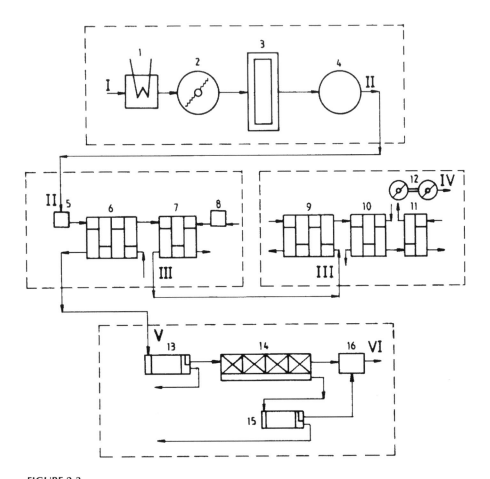

FIGURE 2.3

A schematic flow sheet of the selected operations, processes, and equipment.

Note: 1: Cooler; 2: thickener; 3: vacuum filter; 4: adsorption C; 5: oxidation reactor; 6: four M-S-E units; 7: three M-S-S units; 8: reduction reactor; 9: four M-S-E units; 10: three M-S-W units; 11: two M-S-S units; 12: two crystallizer units; 13: lamellar coalescer; 14: A-I-F cells; 15: settler; and 16: raffinate tank.

Figure 2.4), which was designed on the basis of the laboratory batch and continuous studies [28, 33]. Under the hydrodynamic equilibrium, the index of mixing and power of mixing were measured. During the experiments, the light phase was continuous, and the variable parameters were as follows:

- The ratio of phases in the mixer, light:heavy, was 1.5, 1.7, 2.0, and 2.5.
- The number of revolutions of an eight-blade double-shrouded impeller in the mixer varied in the range of 400–1100 min^{-1}.
- The temperature of the liquid phases in the mixer-settler was 298, 313, and 323 K.

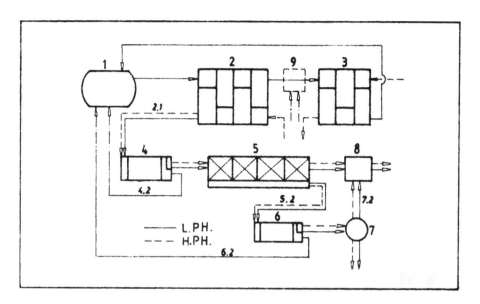

FIGURE 2.4

The light-phase loop.

Note: 1: The light-phase tank; 2: the mixer-settler extraction units; 3: the mixer-settler reduction stripping units; 4: the lamellar coalescer, central unit; 5: the flotation units; 6: the settler unit; 7: the adsorption column unit; 8: the raffinate tank; and 9: the light-phase collector tank.

Figure 2.5 show the graphs, including the experimental points of the correlations: (a) the efficiency of mixing—the number of revolutions of the impeller [$I_m = 1 + \exp(M_m - G_m^3 D_m^2)$, where M_m and G_m were determined by hold up measurements]; (b) the specific power of mixing—the number of revolutions of the impeller [$P = C_m m M_s$, where P was determined by momentum measurements]; and (c) the height of the dispersion band in the settler—the volumetric flow rate density of the polydispersion in the settler [$H_d = K(Q_d/S)^m$ and $H_d = a_s n + b$], respectively.

The selected hydrodynamic characteristics of the mixer-settler unit were as follows:

- The ratio of phases in the mixer, light:heavy, was equal to 1.7.
- The number of revolutions of an eight-blade double-shrouded impeller in the mixer was equal to 11.66 s^{-1}.
- The mixing criterion, $\rho n^3 D^2$, was equal to 27.
- The specific power of mixing, at 313 K, was equal to 3 kW m^{-3}.
- The volumetric flow rate density of the polydispersion in the settler was equal to 0.81 m^3 s^{-1} m^{-2} × 10^{-3}.
- The dispersion band depth in the settler was equal to 0.021 m.

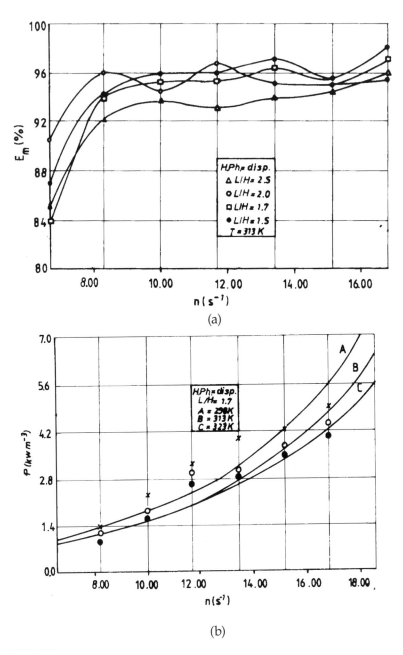

FIGURE 2.5

The selected hydrodynamic characteristics in the mixer and in the settler.

Note: a: The efficiency of mixing—the number of revolutions of the impeller; b: the specific power of mixing—the number of revolutions of the impeller; and c: the height of dispersion band in the settler—the volumetric flow rate density of the polydispersion in the settler.

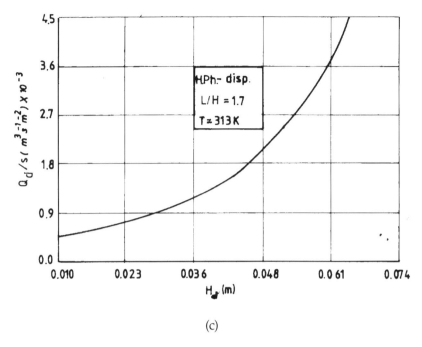

(c)

FIGURE 2.5
(*Continued*)

Figure 2.6 shows an *in situ* photograph of the particular droplet-film struc-tures submerged into the droplet homophase continuum.

The entrained light phase in the form of the double emulsion was led into the lamellar coalescer (numbered 4 in Figure 2.4). Further on, the raffinate was led through four air-induced flotation units (5 in Figure 2.4; Denver A 5 type). The flotation overflow was split in the settler (6), and the heavy-phase fraction was led into the adsorption column (7) filled with activated carbon. Since the separated light phase contained some uranium, collected output was led back into the reductive stripping units (3).

2.1.2 Entrainment Problems in Solvent Extraction: Performance of Demulsions

The volume fractions of the entrained light phase at the input and the out-put of the lamellar coalescer were determined by centrifugal separation in the bottles with capillary tubes on its top; this method was used for the determination of volumetric fractions of the entrained light phase between fifty and several hundreds of parts per million (ppm) [45]. Further on, the volume fractions of the entrained light phase at the input and the output

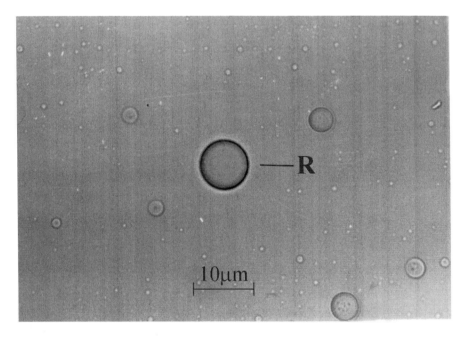

FIGURE 2.6
Photograph *in situ* of the examined liquid-liquid droplet-film structure submerged into the droplet homophase continuum.

of the flotation unit's battery and adsorption column were determined by gas chromatography; volumetric fractions of the entrained light phase found were between 20 and 70 parts per million. By gas chromatography beside the entrained light phase, the dissolved D2EHPA and TOPO in the heavy phase were determined. Since their solubility is negligible (ca. 5–7 ppm vol/vol), the correlation of these methods is possible [45]. The regression analysis of 17 parallel measurements gives the relation $Y = 0.75X + 49.7$ with the correlation coefficient k_r equal to 0.959. The content of the entrained light phase at the input of the lamellar coalescer, at 90% confidence, was $0.0533 + 0.0103\%$ vol/vol; the content of the entrained light phase at the output of the lamellar coalescer, at 90% confidence, was $0.0205 + 0.0035\%$ vol/vol, which was the content of the entrained light phase at the input of the flotation units; and the residual volumetric fraction of the entrained light phase, at the output of flotation units, at 90% confidence, was $0.0053 + 0.0016\%$ vol/vol.

On the basis of the obtained results, the material balance for raffinate and entrained light phase is evaluated and presented in Table 2.1.

The efficiency of the lamellar coalescer unit was 62%. Further on, the efficiency of the flotation units was 77%. The feed fraction in the adsorption column was small; therefore, its efficiency was 63%. Finally, the overall efficiency of the examined loop for the separation of the entrained light phase was 92%.

TABLE 2.1

The Material Balance for the Raffinate and the Entrained Light Phase

Equipment and Position	Position in Figure 2.2	Volume of the Raffinate (dm³)	Distribution of the Raffinate (%)	Fraction of the Light Phase (%)	Volume of the Light Phase (dm³)	Distribution of the Light Phase (%)
Coalescer	4					
Input	2.1	111.0000	100.0000	0.0533	0.059163	100.00
Underflow	4.1	110.9636	99.9672	0.0205	0.022755	38.46
Overflow	4.2	0.0364	0.0328	99.9999	0.036408	61.54
Flotation units	5					
Input	4.1	110.9636	99.9672	0.0205	0.022755	38.46
Underflow	5.1	99.8672	89.9704	0.0040*	0.003994	6.75
Overflow	5.2	11.0963	9.9967	0.1690	0.018760	31.71
Settler	6					
Input	5.2	11.0963	9.9967	0.1690	0.018760	31.71
Underflow	6.1	11.0887	9.8808	0.0100	0.001109	1.88
Overflow	6.2	0.0176	0.0159	99.9999	0.017650	29.83
Adsorption column	7					
Input	6.1	11.0787	9.9808	0.0100	0.001109	1.88
Output 1	7.1	11.0780	9.9801	0.0037	0.000410	0.70
Output 2	7.2	0.0007	0.0007	99.9999	0.000699	1.18

* As the solubility of the D2EHPA and TOPO in H_3PO_4 is approximately equal to 0.0005–0.0007% vol/vol, the value of 0.0053% is decreased to 0.0040%, i.e., for 13 ppm vol/vol.

Determination of the first and the second critical diameters has been performed by an optical microscope *in situ*. The first and the second critical diameters of the examined droplet-film structures are 8 μm and 1 μm, respectively.

2.1.2.1 Mechanism of the Droplet-Film Rupture on an Inclined Plate

Now, the central equipment in the loop relevant to this investigation is the lamellar coalescer, and its schematic cross section, including the design parameters, is presented in Figure 2.7.

The breakage processes are the physical processes initiated by the electromagnetic oscillation that causes the tuning or structuring of the molecules (ions) in the double electrical layer. The structuring is realized by complex motions over the various degrees of freedom whose energy contribution depends on the position of the individual molecules in and around the stopped droplet-film structure on the inclined plate [38]. If the breakage processes are followed along the path of one droplet-film structure from the

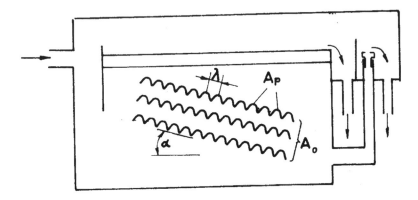

FIGURE 2.7
A schematic cross section and design parameters of the lamellar coalescer.

input in the lamellar coalescer up to its output, then the events occurring can be presented by the series: approach, rest, disturbance, rupture, coalescence (underflow), flow up, and coalescence (overflow).

2.1.2.2 Electromechanical Analogy: Interfacial Tension and Electrical Interfacial Potential

A fundamental approach will be used to analyze a special stability and rupture problem of the droplet-film structure immersed in the droplet homophase continuum. Formation and rupture processes of the secondary liquid-liquid droplet-film structures will be discussed considering mechanical and electrical principles. The analogy of interfacial tension and interfacial electric potential will be illustrated considering the physical model of the processes appearing during the secondary separation of the droplet-film structure submerged in the droplet homophase continuum (double emulsion) on an inclined plate. Figure 2.8 shows the physical model of the processes involved; approach, rest, disturbance, rupture, and flow up.

The generator pole is the origin and source of the disturbance, and the rupture pole is the point where the electrical and mechanical waves change the direction of traveling (feed in and feed back).

The effect of an acting force on the rate of thinning of the film covering the secondary drops is to be discussed. Because of the pressure gradient associated with flow in the film and because a fluid-liquid interface must deform when there is a pressure difference across it, the film thickness varies with position, giving rise to the well-known dimple [53]. An applied or an acting force changes the distribution of the pressure in the film, and hence the variation in its thickness with position and time.

APPROACH   REST   DISTURBANCE RUPTURE FLOW UP

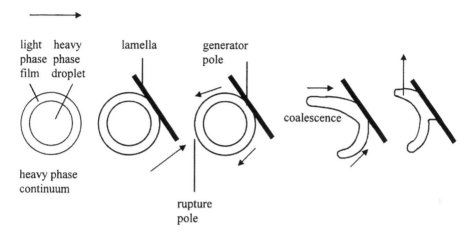

FIGURE 2.8
The physical model of the processes during the secondary separation of the double emulsion on an inclined plate.

The film thickness of a droplet-film structure δ_0, at rest, is calculated from the relation given by the following [51]:

$$\delta_0 = 0.707d \left(\frac{d^2 \Delta \rho g}{\sigma_{in}} \right)^{0.5} \tag{2.1}$$

where d is the droplet-film structure diameter, $\Delta \rho$ is the density difference between the two liquids, σ_{in} is the interfacial tension, and g is the acceleration due to gravity.

Coalescence of the primary liquid drops and gas bubbles with their homophase is determined by the rate of drainage of the intervening fluid film. For a uniform film of viscosity μ, the variation in film thickness δ with time t is given by the following [56]:

$$-\frac{d\delta}{dt} = \frac{8\pi}{3n^2} \frac{\delta^3}{\mu} \frac{f_c}{a^2} \tag{2.2}$$

where f_c is the force pressing on the film of area a, and n is the number of immobile interfaces bounding the film. Thus, in the case of a double emulsion stopped on an inclined plate, n is equal to 1, since there is one immobile interface.

Coalescence of the secondary droplets from the liquid-liquid droplet-film structures with their homophase is limited by the rate of rupture of the covering fluid film. When the droplet-film structure stops on an inclined plate, the film surface is in hydrodynamic equilibrium and the electric double layers are at rest. Since the droplet-film structures are very small, it is postulated that the electrical forces become dominant when compared with the viscous forces on the junction point between the droplet-film structure and the plate. Based on the hypothesis that the electrical forces are responsible for the droplet-film rupture, one can adopt Eq. (2.2) with necessary changes relevant to the nature of forces involved in the rupture process [51]. Now, the variation in film thickness δ with time t is given by

$$-\frac{d\delta}{dt} = \frac{8\pi}{3} \frac{\delta^3}{Z} \frac{f_e}{a^2} \qquad (2.3)$$

where f_e is the electrical force acting on the droplet-film structure surface of area a, and Z is the complex resistance or impedance of the equivalent electric circuit. Therefore, the rate constant in differential Eq. (2.3) is given by

$$C_m = \frac{8\pi}{3} \frac{1}{Z} \frac{f_e}{a^2} \qquad (2.4)$$

Also, the electrical force f_e may be introduced using an analogous relation to

$$\frac{f_c}{a^2} = \frac{2\sigma_{in}}{v} \qquad (2.5)$$

Namely,

$$\frac{f_e}{a^2} = \frac{2\sigma_{ep}}{v} \qquad (2.6)$$

where v is the droplet-film structure volume, and σ_{ep} is the electrical interfacial potential.

2.1.2.2.1 Formation Process

The study of the spreading phenomena and interfacial instability at the surface/interface boundary of two immiscible liquids has been the subject of few publications, in spite of the fact that such information has a great importance for the examination of the rupture mechanism. During the formation of the secondary liquid-liquid structures, viscous and electrical forces are predominantly involved. Therefore, hydrodynamic and electrodynamic equilibrium have to be reached.

Existence of the secondary liquid-liquid droplet-film structure or its electroviscoelastic behavior is dependent on the following:

Droplet film particle size.

Curvature of the droplet-film interface.

Density difference between the phases.

Viscosity ratio of the phases.

Impedance ratio of the phases.

Interfacial tension.

Interfacial electrical potential.

Temperature effects.

Mechanical effects (vibration).

Third phase presence.

Mutual solubility.

External periodical physical fields (e.g., temperature, electric, magnetic, and ultrasonic).

Internal periodical physical fields (mechanical, electric, and magnetic).

2.1.2.2.2 Stability

Figure 2.9 shows the graphical interpretation of a droplet-film structure stopped on the inclined plate, and an acting force with its components at the structure-plate junction point.

Figure 2.10 shows the graphical interpretation of a droplet-film structure approach to the inclined plate, equilibrium and rupture using mechanical principles. F_x, F_y, and F_z are the component vectors of F_s, which is the resultant surface force vector in 3N dimensional configuration space.

According to Newton's second law, the general equation of fluid dynamics in differential form is given by

$$\rho \frac{D\tilde{u}}{Dt} = \sum_i \tilde{F}_i (dxdydz) + d\tilde{F}_s \qquad (2.7)$$

When a droplet-film structure rests on the inclined plate, the term on the left-hand side of Eq. (2.7) becomes equal to zero; further, the terms on the right-hand side represent the volume, F_i (gravitational F_g, buoyancy F_{bo}, and electromagnetic/Lorentz F_l), and the surface, F_s forces, respectively. The gravitational force is superimposed to the buoyancy force; therefore, the volume force term is equal to zero. The surface forces are supposed to be associated to the interface between the fluid continuum and the droplet, and between the latter and the plate. They can be calculated for by means of a tensor T_n as follows:

$$d\tilde{F}_s = \tilde{T}_n d\tilde{A} \qquad (2.8)$$

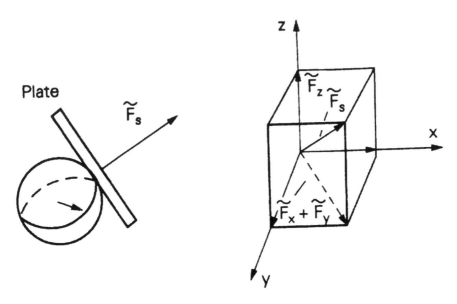

FIGURE 2.9

A graphical interpretation of a droplet-film structure-plate junction point.

Source: Spasic, A. M., Mechanism of the secondary liquid-liquid droplet-film rupture on inclined plate, *Chem. Eng. Sci.*, 47, 1992, p. 3952, with permission from Elsevier Science.

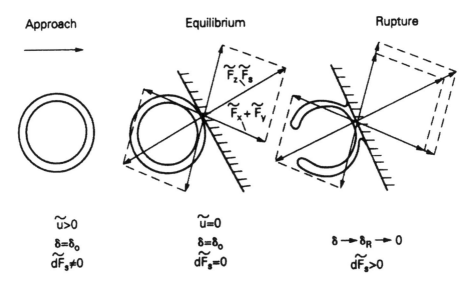

FIGURE 2.10

Mechanical principles, the droplet-film structure approach, equilibrium, and rupture.

Source: Spasic, A. M., Mechanism of the secondary liquid-liquid droplet-film rupture on inclined plate, *Chem. Eng. Sci.*, 47, 1992, p. 3953, with permission from Elsevier Science.

where T_n is composed of two tensors given by

$$\tilde{T}_n = -p\left[\delta_j^i\right] + \left[\xi_j^i\right] \tag{2.9}$$

In the first, δ_j^i is the Kronecker symbol, isotropic tensor the hydrostatic pressure is dominant and the contribution of the other elements is neglected; further, in the second stretching tensor ζ_j^i the tangential elements are presumed to be of adhesive origin, and the normal elements are due to the interfacial tensions.

The force balance at the junction point between the plate and the droplet-film structure is given by the following:

$$d\tilde{F}_s - \left(d\tilde{F}_x + d\tilde{F}_y + d\tilde{F}_z\right) = 0 \tag{2.10}$$

The tangential elements τ in the second stretching tensor of Eq. (2.9) are presumed to be identical; therefore, the mechanical equilibrium condition is given by

$$\tau = \frac{p - \left(\frac{\sigma_{in}}{d}\right)}{2} \tag{2.11}$$

Introducing the impedance Z instead of the viscosity μ and the electron flux density ϕ instead of the velocity u, an electrical analogous approach to the droplet-film rupture mechanism on the inclined plate can be proposed, as shown in Figure 2.11.

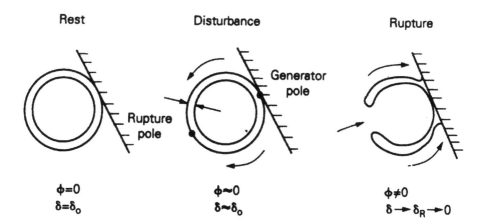

FIGURE 2.11
Electrical principles, a droplet-film structure rest, disturbance, and rupture.

Source: Reprinted from Spasic, A. M., Mechanism of the secondary liquid-liquid droplet-film rupture on inclined plate, *Chem. Eng. Sci.*, 47, 1992, p. 3953, with permission from Elsevier Science.

At first, the droplet-film structure surface is at rest, and then the structure-plate junction point becomes the generator pole (source or sink of the incident sinusoidal wave) and disturbs the electric double layer changing the distribution of pressure in the film and, hence, causing the variation in the film thickness, which ends with the rupture. Now, the analogous electrical equilibrium condition is given by

$$\tau = \frac{U - \left(\frac{\sigma_{ep}}{d}\right)}{2} \tag{2.12}$$

where U is the electrostatic potential and σ_{ep} is the electrical interfacial potential.

For spherical polydispersions, the relation between the interfacial tension σ_{in} and the internal pressure p_{in} in the droplet-film structure is given by Eq. (2.13):

$$\frac{\sigma_{in}}{d} \approx p_{in} \tag{2.13}$$

and the relation between the electrical interfacial potential σ_{ep} and the electrical internal potential U is given by

$$\frac{\sigma_{ep}}{d} \approx U \tag{2.14}$$

So, a discussion of the formation, hydrodynamic, and electrodynamic equilibrium as well as rupture processes is facilitated in terms of the proposed electromechanical analogy.

2.1.2.2.3 Rupture Process

The rupture process is to be analyzed for the special case of the droplet-film structure on an inclined plate. Now, under the assumption that about the same interfacial jump potential appears during the droplet-film formation and rupture processes, it is postulated that the generator pole (junction point of the droplet-film structure and the inclined plate) is the source or sink of the incident sinusoidal wave. Hence, the impedance of the structure consists of resistance and reactance terms. The surface of the droplet-film structure resting on the inclined plate is supposed to achieve initially a hydrodynamic and electrodynamic equilibrium. Therefore, from Eq. (2.3) the process of the film thinning is given by

$$\delta = \frac{\delta_0}{\left(1 + 2\delta_0^2 C_m t\right)^{0.5}} \tag{2.15}$$

where C_m is given by Eq. (2.4). When time tends to the rupture time t_R, the film thickness δ tends to zero, which may be represented by

$$\lim_{t \to t_R} \delta = \delta_R \to 0 \tag{2.16}$$

In general, after the rupture process of the light-phase film is completed, the coalescence process of the heavy-phase droplet with its homophase occurs. Factors affecting the secondary coalescence time are as follows [3]:

Droplet-film structure size.

Distance of fall of droplet to the interface.

Curvature of the droplet-side interface.

Density difference between phases.

Viscosity ratio of the phases.

Interfacial tension effects.

Temperature effects.

Vibration and electrical effects.

Presence of the electrical double layers.

Solute transfer effects.

It can be seen in Figure 2.8 that the phases of the overall process during the secondary separation of a double emulsion on an inclined plate are approach, rest, disturbance, rupture, and flow up. Besides the analysis of the wave propagation or disturbance spreading by the impedance model, the secondary separation process can be represented by successive time sequences as follows:

$$t_{SE} = t_A + t_{RE} + t_{DIST} + t_R + t_{FU} \qquad (2.17)$$

Under the assumption that the rest t_{RE} and rupture t_R times are infinitely short, and since the two electric double layers are to be destroyed, the disturbance time t_{DIST} is composed of two disturbance and two collapse subsequences, and may be written as

$$t_{DIST} = t_{distII} + t_{colII} + t_{distI} + t_{coll} \qquad (2.18)$$

Hence, for this particular case after the rupture process is completed, the start of "flow up" (lifting of the separated light-phase film) takes place. Again, the volume forces (gravitational, buoyancy, and electromagnetic) enter into the game [51,52]. The flow up occurs when the following condition is satisfied:

$$\tilde{F}_{bo}dxdydz + \tilde{F}_l dxdydz \geq \tilde{F}_g dxdydz \qquad (2.19)$$

2.1.2.3 Designing of the Lamellar Coalescer

A method for predicting the performance of demulsion involves the relationship between the phase separation operation or method and the secondary liquid-liquid contact system, taking into account limitations of the equipment or plant capacity, and maximal permitted content of residual emulsion or wanted sensitivity. This work conveys the knowledge organization or process synthesis, the main objectives of which are to enable a choice and enable designing of the phase separation method, equipment, and/or plant conception [57,67]. The particular problem considered in this chapter is the process synthesis in the treatment of entrainment in liquid-liquid extraction. The experimental results obtained in the pilot plant for uranium recovery from wet phosphoric acid were used as the comparable source [45,67].

2.1.2.3.1 *Necessary Mathematical Models*

These models are the movement model, levitation model, impedance model, rupture model, and buoyancy model.

According to Newton's second law, the general equation of fluid dynamics in differential form is given by

$$\rho \frac{D\tilde{u}}{dt} = \sum_i \tilde{F}_i(dxdydz) + d\tilde{F}_s \tag{2.20}$$

Equation (2.20) covers the approach period of a droplet-film structure toward the inclined plate, what may represent the *movement model*. Further on, when a droplet-film structure stops on the inclined plate, the term on the left-hand side of Eq. (2.20) becomes equal to zero. The terms on the right-hand side represent the volume and the surface forces. The gravitational force is superposed with the buoyancy and electromagnetic (Lorenz) forces; therefore, the volume forces' term is equal to zero too, which may represent the "forced" *levitation model* written as

$$(\tilde{F}_g + \tilde{F}_b + \tilde{F}_l)dxdydz = 0 \tag{2.21}$$

The surface forces are supposed to be composed of the fluid continuum-droplet-film structure surface, and the plate-droplet-film structure surface interaction terms given by

$$d\tilde{F}_s = \tilde{T}_n d\tilde{A} \tag{2.22}$$

where the tensor T_n is expressed by

$$T^{ij} = -\alpha_0 \delta^{ij} + \alpha_1 \delta^{ij} + \alpha_2 \zeta^{ij} + \alpha_3 \zeta_k^i \zeta^{kj} \tag{2.23}$$

In the first isotropic tensor, the potentiostatic pressure $\alpha_0 = \alpha_0(\rho, U)$ is dominant and the contribution of the other elements is neglected. In the second isotropic tensor, the resistance $\alpha_1 = \alpha_1(\rho, U)$ is dominant and the contribution of the other elements is neglected. In the third stretching tensor, normal elements $\alpha_2 \sigma$ are due to the interfacial tensions, and the tangential elements $\alpha_2 \tau$ are presumed to be of the same origin as the dominant physical field involved. In the fourth stretching tensor, there are the normal $\alpha_3 \sigma_k^i$ and $\alpha_3 \sigma^{kj}$

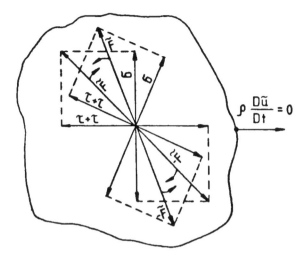

FIGURE 2.12
Equilibrium of the surface forces at any point of a stopped droplet-film structure while in an interaction with some periodical physical field (a two-dimensional projection).

Source: Reprinted from Spasic, A. M., Jokanovic, V., Krstic, D. N., A theory of electroviscoelasticity: A new approach for quantifying the behavior of liquid-liquid interfaces under applied fields, *J. Colloid Interface Sci.*, 186, 1997, p. 438, with permission from Academic Press.

elements, and the tangential $\alpha_3 \tau_k{}^i$ and $\alpha_3 \tau^{kj}$ elements that are attributed to the first two dominant periodical physical fields involved. Now, the general equilibrium condition for the dispersed system with two periodical physical fields involved may be derived from Eq. (2.23), and may be expressed by

$$\tau_{dom} = \frac{-\alpha_0 + \alpha_1 + \alpha_2\left(\frac{\delta}{d}\right) + \alpha_3\left(\frac{\delta}{d}\right)}{2(\alpha_2 + \alpha_3)} \tag{2.24}$$

where τ_{dom} is the tangential stress of the same origin as that of the dominant periodical physical field involved. Figure 2.12 shows the schematic equilibrium of surface forces at any point of a droplet-film structure while in an interaction with some periodical physical field, for example, at the inclined plate.

Taking into consideration all established assumptions, and introducing the impedance Z instead of the viscosity μ and the electron flux density ϕ_e instead of the velocity u, the electrical analogous approach to the droplet-film rupture mechanism on the inclined plate is shown in Figure 2.11. At first, the droplet-film structure surface is at rest; then, the structure-plate junction point becomes the generator pole (source or sink of the incident sinusoidal wave) and disturbs the electrical double layer, changing the distribution of pressure in the film and, hence, causing the variation in the film thickness

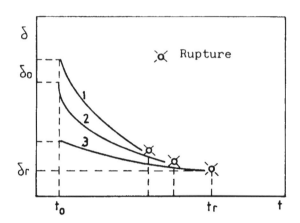

FIGURE 2.13

Solutions of the film-thinning process equation for different droplet-film structure diameters.

Note: (1) The first critical diameter; (2) the representative, arbitrarily chosen diameter; and (3) the second critical diameter.

which ends with the rupture. When the disturbance period begins, the electromagnetic (Lorenz) forces cause the complex motion over the various degrees of freedom. At first, Lorenz forces initiate "breathing" of the droplet-film structure which may, depending on the energy contribution, cause its rupture or plastic deformation. The variation in film thickness with time is governed by the *impedance model*, and this relation is given by

$$-\frac{d\delta}{dt} = \frac{8\pi}{3} \frac{\delta^3}{Z} \frac{f_e}{a^2} \tag{2.25}$$

where δ is the thickness of the film, Z is the impedance, and f_e is the electrical force acting on the surface a. Figure 2.13 shows the solutions of the film-thinning process, Eq. (2.25), for different droplet-film structure diameters.

Now, besides the wave propagation or disturbance analysis by the impedance model, the rupture process can be presented by successive time sequences where its structure may be studied. The secondary coalescence time may be presented by the *rupture model* that is given by

$$t_{co}''' = t_{re} + t_{dist} + t_{ob} \tag{2.26}$$

where t_{re} is the rest time or the past; t_{dist} is the disturbance time or the present, which is composed of two disturbance and two collapse time subsequences, and may be written as

$$t_{dist} = t_{dist2} + t_{cil2} + t_{dist1} + t_{col1} \tag{2.27}$$

and t_{ob} is the initial buoyancy time or the future. Under the assumption that the rest and initial buoyancy times are infinitely short, the secondary coalescence time or the rupture time is circa 500 ms. Here, again, the volume forces enter into the game, and the *buoyancy model* is presented by

$$(\tilde{F}_b + \tilde{F}_l)dxdydz > \tilde{F}_g \forall \qquad (2.28)$$

2.1.2.3.2 Overall Volume Balance Model

The lamellar coalescer as a central equipment in the entrained light-phase separation loop, formerly shown, may be understood as a steady-state plug flow reactor under the assumption that the droplet-film structure is the reactant. Figure 2.7 shows a schematic cross section and the design parameters of the lamellar coalescer. The capacity parameters are the overall area of the filling plates A_0, and the overall area of the perforations on the top of the corrugated plates A_p. Further on, the semimacro parameters are the inclination angle of the perforated corrugated plates α, and the distance between two consecutive tops of the plate λ. Finally, the process parameter or molecular parameter is the wetting angle between the light phase and the filling plate material γ.

According to the main assumption of the model, the differential equation governing the variation of the entrained light-phase volume V with time t in the lamellar coalescer is given by

$$-\frac{dV}{dt} = K_{or}V \qquad (2.29)$$

The boundary conditions, according to the results obtained in the separation loop, for the throughput of 30.5 m³s⁻¹× 10⁻⁶ and for the residence time of 960 s (at 90% confidence, input 0.0533 + 0.0103% vol/vol, and output 0.0205 + 0.0035% vol/vol) are

$$t_1 = t_0 = 0 \ s; \quad V_1 = 0.0591 \ \text{m}^3 \times 10^{-3}$$

and

$$t_2 = t_R = 960 \ s; \quad V_2 = 0.0227 \ \text{m}^3 \times 10^{-3}$$

The integral of Eq. (2.29) is given by

$$V = K_{or} \exp(-K_{or}t) \qquad (2.30)$$

The overall rate constant is a function of the capacity, semimacro, and molecular parameters, which may be written as

$$K_{or} = f[(A_0, A_p), (\alpha, \lambda), \gamma] \qquad (2.31)$$

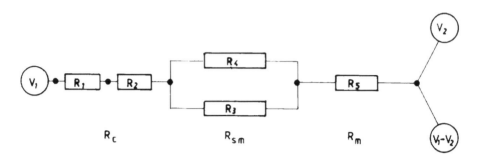

FIGURE 2.14
The structure of the overall rate constant K_{or}.

The structure of the overall rate constant K_{or} is another major assumption of the model, which is represented as a succession of resistances as shown in Figure 2.14.

Hence, the overall coalescence process in the lamellar coalescer is composed of the successive events that correspond to the adequate resistances. The addition of these resistances is assumed considering the successive time sequences and the parameter influence on the separation process. Therefore, the overall rate constant K_{or} is given by

$$\frac{1}{K_{or}} = R_c + R_{sm} + R_m \tag{2.32}$$

The calculated value of the overall rate constant K_{or} is equal to 0.001 s^{-1}, with the values of the particular rate constants K_{oc}, K_{osm}, and K_{om} equal to 0.008 s^{-1}, 0.03 s^{-1}, and 0.0013 s^{-1}, respectively.

Further on, the equivalent resistance is given by

$$\frac{1}{K_{or}} = R_e \tag{2.33}$$

From Figure 2.14, the equivalent resistance is given by

$$R_e = R_1 + R_2 + \frac{1}{\frac{1}{R_3} + \frac{1}{R_4}} + R_5 \tag{2.34}$$

where the resistances R_1 and R_2 correspond to the first event limited by the capacity parameters, the area of the filling plates A_0, and to the last event limited by the area of the perforations on the top of these plates A_p, respectively. Further on, the resistances R_3 and R_4 correspond to the events limited by semimacro parameters, the inclination angle of the filling plates α, and to the distance between two consecutive tops on the filling plate λ (these

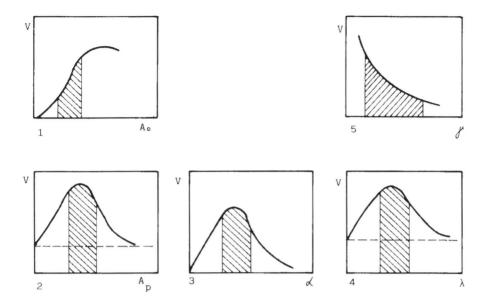

FIGURE 2.15
The solutions of the entrained light-phase balance equation.

resistances act simultaneously), respectively. Finally, the resistance R_5 corresponds to the event limited by the molecular parameter, the wetting angle between the light phase and the filling plate material γ. Now, the entrained light-phase balance equation includes the influence of the capacity semi-macro and molecular parameters, and considering the main assumption of the model presented in Eq. (2.29) is given by

$$V_1 - V_2 = \sum_{i=1}^{5} k_i (parameter_i)^{b_i} \exp[K_{ori}(parameter_i)] \qquad (2.35)$$

Figure 2.15 shows the solutions of Eq. (2.35) regarding different parameters and the k, b, and K_{or} values.

Figure 2.15 shows (at the part of the figure numbered 1) the influence of the overall area of the filling plates A_0 on the volume of the separated light phase V. It can be seen, at the inflection point, that the lamellar coalescer degenerates into the densely packed coalescer (the rupture mechanism is changing). Furthermore, there is no need to increase the overall area A_o after a certain value, because the efficiency is rising slowly while maintenance problems rise as well as the equipment price. Figure 2.15 (at 2) shows the influence of the overall area of the perforations of the filling plates A_p on the separated light-phase volume V. The perforations on the tops of the filling plates permit the "flow up" of the separated light phase. It can be seen that the efficiency is about the same, for the limiting cases, when the plates are either not or very perforated. Figure 2.15 (3) shows the influence of the inclination angle of the filling plates α on the volume of the separated light phase V. The inclination angle α, together with the distance between two consecutive tops on the filling plate λ, enables droplet-film structures to stop at the inclined plate in the convenient position (forced levitation). Figure 2.15 (4) shows the influence of the distance between two consecutive tops of the filling plate λ on the volume of the separated light phase V. It can be seen that the efficiency is about the same, for the limiting cases, when the plates are either plane or very corrugated. Figure 2.15 (5) shows the influence of the wetting angle between the filling plates and the light-phase film γ on the separated light-phase volume V. It can be seen that two contradicting conditions have to be satisfied: not only the adhesion forces, responsible for the droplet-film rupture, have to be strong enough but also the "flow up" of the light phase has to be possible.

2.1.2.3.3 Reduced Semiempirical Mathematical Model

If the overall area of the filling plates A_0 is arbitrarily determined and the plate material, wetting angle γ, is chosen, then the reduced semiempirical mathematical model can be derived and solved. Now, the balance Eq. (2.35) in the reduced form is

$$V_1 - V_2 = \sum_{i=1}^{3} k_i (parameter_i)^{b_i} \exp(-K_{ori}t) \tag{2.36}$$

where the constants K_{ori} and K_{or} according to Eqs. (2.30) and (2.35), are given by

$$K_{or1} = \sum_{i=1}^{3} k_i (parameter_i)^{b_i} \tag{2.37}$$

and the particular rate constants are given by

$$-K_{ori}t = K_i (parameter_i) \tag{2.38}$$

Further, using Eqs. (2.35) and (2.38), the particular resistances are evaluated and are

$$R_1 = \frac{t}{K_1 A} = K_{fr1} ; \text{constant} \tag{2.39}$$

$$R_2 = \frac{t}{K_2 A_p} = R_2 ; \quad A_p \in (0.5 - 2.0\% \ A) \tag{2.40}$$

$$R_3 = \frac{t}{K_3 \alpha} = R_3; \quad \alpha \in (0.17 - 0.34 \text{ rad}) \tag{2.41}$$

$$R_4 = \frac{t}{K_4 \lambda} = R_4; \quad \lambda \in (0.01 - 0.06 \text{ m}) \tag{2.42}$$

$$R_5 = \frac{t}{K_5 \gamma} = K_{fr5}; \text{ constant} \tag{2.43}$$

The resistances R_1 and R_5 are fixed; therefore, their values become constants. Thereafter, the equivalent overall resistance is evaluated from Eq. (2.34) and is given

$$R_e = \frac{K_{fr1} R_3 + K_{fr1} R_4 + R_2 R_3 + R_2 R_4 + R_3 R_4 + K_{fr5} R_3 + K_{fr5} R_4}{R_3 + R_4} \tag{2.44}$$

Now, the overall balance equation for the separated light phase in the lamellar coalescer is given by

$$V = \sum_{i=1}^{3} k_i (parameter_i)^{b_i} \exp(-K_{or} t) \tag{2.45}$$

where the overall rate constant K_{or} is evaluated from Eqs. (2.33) and (2.44). Hence the total differential of Eq. (2.45) may be written as

$$dV = \left(\frac{\partial V}{\partial A_p}\right)_{\alpha, \lambda} dA_p + \left(\frac{\partial V}{\partial \alpha}\right)_{A_p, \lambda} d\alpha + \left(\frac{\partial V}{\partial \lambda}\right)_{A_p, \alpha} d\lambda \tag{2.46}$$

where

$$\left(\frac{\partial V}{\partial A_p}\right)_{\alpha, \lambda} = k_1 A_p^{b_1} \exp(-K_{or} t) \left[\frac{b_1}{A_p} - t \frac{\partial K_{or}}{\partial A_p}\right] \tag{2.47}$$

$$\left(\frac{\partial V}{\partial \alpha}\right)_{A_p, \alpha} = k_2 \alpha^{b_2} \exp(-K_{or} t) \left[\frac{b_2}{\alpha} - t \frac{\partial K_{or}}{\partial \alpha}\right] \tag{2.48}$$

$$\left(\frac{\partial V}{\partial \lambda}\right)_{A_p, \alpha} = k_3 \lambda^{b_3} \exp(-K_{or} t) \left[\frac{b_3}{\lambda} - t \frac{\partial K_{or}}{\partial \lambda}\right] \tag{2.49}$$

The partial derivatives of the overall rate constant K_{or} from Eqs. (2.47)–(2.49), together with the terms in parentheses, are equalized with zero, and when evaluated become three algebraic equations written as

$$a_1 A_p^2 + a_2 A_p + a_3 = 0 \tag{2.50}$$

$$a_4 \alpha^2 + a_5 \alpha + a_6 = 0 \tag{2.51}$$

$$a_7 \lambda^2 + a_8 \lambda + a_9 = 0 \tag{2.52}$$

Finally, the values of the parameters A_p, α, and λ are 1% A_0, 0.25 rad, and 0.03 m, respectively. Only when two of the parameters are determined may this model be solved, for example, the overall area of the filling plates A_0 is arbitrarily determined, and the plate material, wetting angle γ, is chosen. The boundaries, where design parameters can be searched by a developed mathematical model, are based on the consulted literature data as well as on the data obtained from the pilot plant experiments. The parameters of an empirically designed and applied lamellar coalescer together with the boundaries where they can be searched are given by the following:

- $A_0 = 30\ \%\ V_{co}$, from the range; $A_0 \in (20–80\%\ V_{co})$, arbitrary.
- $A_p = 1\ \%\ A_0$, from the range; $A_p \in (0.5–4.0\%\ A_0)$, experimentally.
- $\alpha = 0.25$, rad from the range; $\alpha \in (0.17–0.34\ \text{rad})$, experimentally.
- $\lambda = 0.03$ m, from the range; $\lambda \in (0.01–0.06\ \text{m})$, experimentally.
- $\gamma = ?$, (chosen plate material: poly-methyl-α-methyl acrilat), from the range.
- $\gamma \in =$ (depends on the interaction: filling plate material—the particular liquid-liquid contact system).

2.1.2.3.4 Simple Balance Model

The reduced balance equation is given by

$$\frac{\partial}{\partial t}[n, f_n(V, t)] = B_{co}(V, t) - S_{co}(V, t) \tag{2.53}$$

where the birth function B_{co} and the surviving function S_{co} have to be determined.

Figure 2.16 shows the construction elements of the representative mean volume of the film, the new representative hypothetical monodispersed cumulative droplet-film structure size distribution, and the real overall volume variation of the light-phase films with time in the lamellar coalescer.

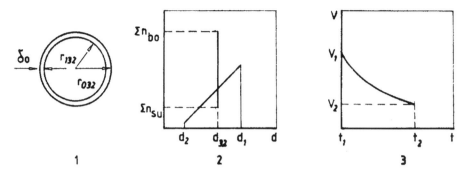

FIGURE 2.16
The construction elements.

Note: 1: The representative mean volume of the film; 2: the representative droplet-film structure size distribution; and 3: the overall volume variation with time of the separated light phase in the lamellar coalescer.

The film thickness δ is used to calculate the representative mean volume V. The film thickness of the generated droplet-film structure δ_0 is calculated from the following relation [51,56]:

$$\delta_0 = 0.707d\left(\frac{d^2\Delta\rho g}{\sigma_{in}}\right)^{0.5}, \tag{2.54}$$

where d is the droplet-film structure diameter, $\Delta\rho$ is the density difference of two liquids, σ_{in} is the interfacial tension, and g is the acceleration due to gravity. Therefore, the representative mean volume of the light-phase film is given by

$$V = V_{032} - V_{132} \tag{2.55}$$

Regarding all established assumptions together with Eqs. (2.30) and (2.55), the birth function may be written as

$$B_{co}(V,t) = n_{bo}\left[V\exp(-K_{or}t)_{t=t_0}^{t=t_R}\right] \tag{2.56}$$

and the surviving function is given by

$$S_{co}(V,t) = n_{su}\left[V\exp(-K_{or}t)_{t=t_0}^{t=t_R}\right] \tag{2.57}$$

The condition that the rupture frequency is equal to zero when the absolute value of time is equal with the residence time of the double emulsion in the lamellar coalescer permits the evaluation of the rupture frequency dependence with time as follows:

$$\omega = C_\omega[\exp(-K_{or}t) - 0.35] \tag{2.58}$$

Finally, the efficiency of the double emulsion breakage in the lamellar coalescer is given by

$$E = \frac{B_{co} - S_{co}}{B_{co}} \qquad (2.59)$$

and

$$E = K_{or} \int_{t_0}^{t_R} \exp(-K_{or}t)dt \qquad (2.60)$$

2.1.3 Marangoni Instabilities of the First and Second Order and a Possible Electrical Analog

In a number of papers on Marangoni instability or the mechanism of Benard cell formation, it is supposed that the surface tension is a linear, monotonically decreasing function of temperature [16,17]. This behavior is typical for a large class of fluids, for example, water, silicone oil, and water benzene solutions. There are exceptions, such as some alloys, molten salts, and liquid crystals, that show a linear growth of surface tension with temperature. Also, there exists a third class of fluid systems characterized by a surface tension showing a nonlinear dependence with respect to temperature. This behavior is representative of aqueous long-chain alcohol solutions and some binary metallic alloys [17].

The Marangoni instability of the first order was first elucidated and demonstrated theoretically in Probstein [16]. It was shown that if there was an adverse temperature gradient of sufficient magnitude across a thin liquid film with a free surface, such a layer could become unstable and lead to cellular convection. This instability mechanism is illustrated in Figure 2.17.

A small disturbance is assumed to cause the film of initially uniform thickness to be heated locally at a point on the surface. This results in a decreased surface tension and a surface tension gradient that leads to an induced motion tangential to the surface away from the point of local heating. From mass conservation, this motion in turn induces a motion of the bulk phase toward the surface. The liquid coming from the heated region is warmer than the liquid-gas interface. The motion is thus reinforced, creating cellular convection patterns, and will be maintained if the convection overcomes viscous shear and heat diffusivity [16]. It is appropriate to introduce the critical parameter termed the *Marangoni number* given by

$$Ma = \frac{\sigma_T \beta_T \delta^2}{\mu} \qquad (2.61)$$

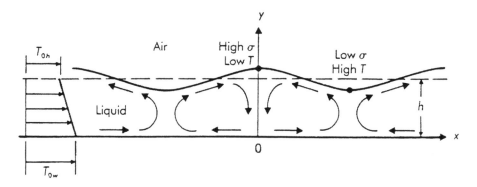

FIGURE 2.17
A Marangoni instability mechanism.

Source: Reprinted from Probstein, R. F., *Physicochemical Hydrodynamics*, Wiley, New York, 1994, p. 354, with permission of John Wiley & Sons, Inc.

where σ_T is the surface tension, β_T is the uniform temperature gradient, δ is the film thickness, and μ is the viscosity. This number can also be interpreted as a thermal Peclet number that represents a measure of the ratio between heat transport by convection due to surface tension gradients, and bulk heat transport by conduction. More detailed discussion on the derivation of the Marangoni number can be found in numerous references [e.g., 16, 17].

Now, it is possible to redefine the parameters with the changes relevant to the nature of forces involved, and to obtain the relation for the electrical critical parameter given by

$$AD_R = \frac{\sigma_E \gamma_T \delta^2}{Z} \tag{2.62}$$

where σ_E is the electrical interfacial potential, γ_T is the uniform potential gradient, δ is the film thickness, and Z is the impedance [11]. The subscript R indicates a rigid form of "instability," namely, to the stable existence of droplet or droplet-film structures. This instability mechanism is illustrated in Figure 2.18a.

Using the analogous approach, where the buoyancy forces are neglected as in, for example, a microgravity environment, a nonlinear dependence of the surface tension with respect to the temperature can give rise to the second-order Marangoni effect, with the characteristic number:

$$Ma'' = \frac{\left(\frac{\partial^2 \sigma_T}{\partial T^2}\right)(\Delta T)^2 \delta}{\rho v k} \tag{2.63}$$

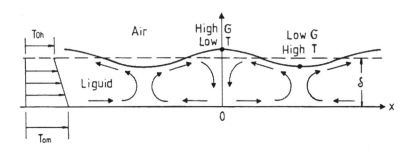

$$Ma = \frac{\sigma_T \beta_T \delta^2}{\mu}$$

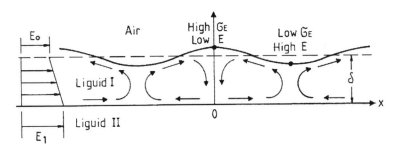

$$AD_R = \frac{\sigma_E \gamma_T \delta^2}{Z}$$

(a)

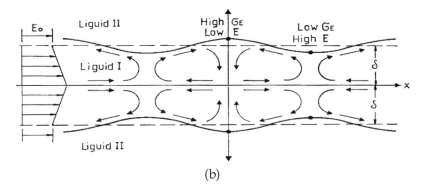

(b)

FIGURE 2.18
A Marangoni instability mechanism: (a) electrical analog, and (b) possible particular situation (e.g., for double emulsion and/or a biological cell).

where ΔT is the temperature drop between the lower and upper boundaries of the layer, δ is the distance between the boundaries of the layer, ρ is the constant density, k is the heat diffusivity, and v is the kinematic viscosity [17].

Finally, the nonlinear electrical critical parameter is derived and expressed by

$$AD_E = \frac{\left(\frac{\partial^2 \sigma_E}{\partial y^2}\right)(\Delta y)^2 \delta}{Z} \tag{2.64}$$

where Δy is the potential drop between the lower and upper boundaries of the layer, δ is the distance between the boundaries, and Z is the impedance. The subscript E corresponds to an elastic form of "instability," indicating formation, breathing, and destruction of a droplet or a droplet-film structure. It is suggested to consult Drexler [7] for other nonlinear interactions.

Now imagine the case of, for example, double emulsion and/or a biological cell; then the situation in Figure 2.18b may be expected only for certain forces in the game, although there may be several possible cases, such as droplet rotation, droplet go-back rotation, film rotation, quasi-equilibrium, or no rotation at all.

2.1.4 Rheology: Various Constitutive Models of Liquids

2.1.4.1 Structure and Dynamics: Various Constitutive Models of Liquids

In the last two centuries, a lot of attempts and discussions have been performed for the elucidation and development of the various constitutive models of liquids. Some of the theoretical models that can be mentioned here are Boltzmann's, Maxwell's (UCM, LCM, COM, IPM), Voigt or Kelvin's, Jeffrey's, Reiner-Rivelin's, Newton's, Oldroyd's, Giesekus's, graded fluids, composite fluids, and retarded fluids with a strong backbone and fading memory [66]. Several of these models will be discussed in some detail, for example, Maxwell's, Voigt or Kelvin's, and Jeffrey's, respectively. Further and deeper knowledge related to the physical and mathematical consequences of the structural models of liquids and of the elasticity of liquids can be found in the book *Fluid Dynamics of Viscoelastic Liquids* by Joseph [66].

2.1.4.2 Maxwell's Model

This model consists of a spring and dashpot connected in series. The spring presents an elastic element, and the dashpot is a viscous element. Figure 2.19a shows the circuit, where G is the spring constant, σ is the force, μ is the viscosity, γ is the displacement of the spring, and $\partial \gamma / \partial t$ is the velocity.

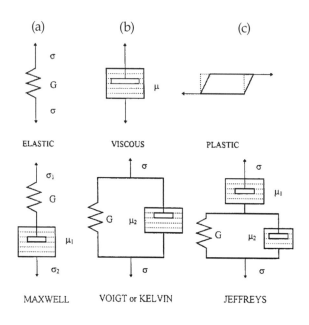

FIGURE 2.19
The rheological models: (a) Maxwell's, (b) Voigt or Kelvin's, and (c) Jeffrey's.

The force, because the elements are in series, is given by

$$\sigma_1 = \sigma_2 = \sigma \qquad (2.65)$$

where

$$\sigma_1 = G\gamma_1 \qquad (2.66)$$

and

$$\sigma_2 = \mu \left(\frac{\partial \gamma}{\partial t} \right)_2 \qquad (2.67)$$

Further on, the velocity or the time rate of change of γ_2, also the time rate of change of the total displacement, is

$$\frac{\partial \gamma}{\partial t} = \left(\frac{\partial \gamma}{\partial t}\right)_1 + \left(\frac{\partial \gamma}{\partial t}\right)_2 = \frac{\left(\frac{\partial \sigma}{\partial t}\right)_1}{G} + \frac{\sigma_2}{\mu} = \frac{\frac{\partial \sigma}{\partial t}}{G} + \frac{\sigma}{t} \tag{2.68}$$

When arranged, it can be written as

$$\lambda \frac{\partial \sigma}{\partial t} + \sigma = \mu \frac{\partial \gamma}{\partial t} \tag{2.69}$$

where the relaxation time λ is given by

$$\lambda = \frac{\mu}{G} \tag{2.70}$$

Another expression for σ is given by

$$\sigma = \frac{\mu}{\lambda} \int_{-\infty}^{t} \exp\left[\frac{-(t-\tau)}{\lambda}\right] \frac{\partial \gamma(\tau)}{\partial \tau} d\tau \tag{2.71}$$

Obviously, Eq. (2.69) is a differential equation model for the relation between force and deformation, and Eq. (2.71) is an integral model showing that the present value of the force $\sigma(t)$ is determined by the history of γ.

Equation (2.69) is the constitutive equation for Maxwell's model. For a displacement $\gamma \neq 0$, and $\partial \gamma / \partial t = 0$; it follows from Eq. (2.71) that

$$\sigma = \sigma(0) \exp\left(-\frac{t}{\lambda}\right) \tag{2.72}$$

At $t = 0$, when displacement γ increases, the production of stress $\sigma(x,t)$ occurs; and when displacement γ is constant, then the stress relaxes, eventually to zero. Therefore, Maxwell's circuit has instantaneous elasticity. Fluids with long times of relaxation are nearly elastic, and those with short times of relaxation are very nearly viscous. Hence, discussion related to the relaxation time λ leads to the following consequences: for $\lambda \rightarrow 0$, the Newtonian fluid follows, thus

$$\sigma = \mu \frac{\partial u}{\partial x} \tag{2.73}$$

and

$$\rho \frac{\partial u}{\partial t} = \mu \frac{\partial^2 u}{\partial x^2} \tag{2.74}$$

This is a parabolic problem that smoothes discontinuities by diffusion. For $\lambda \to \infty$, the elastic fluid follows and

$$\frac{\partial \sigma}{\partial t} = G \frac{\partial u}{\partial x} \tag{2.75}$$

or

$$\sigma = G \frac{\partial \xi}{\partial x} \tag{2.76}$$

This is a hyperbolic equation that allows wave propagation of discontinuities without smoothing. Also, the telegraph equation is given by

$$\frac{\partial^2 u}{\partial \tau^2} + \frac{1}{\lambda} \frac{\partial u}{\partial t} = \frac{\mu}{\partial \lambda} \frac{\partial^2 u}{\partial x^2} \tag{2.77}$$

This is an equation of the type

$$\frac{\partial^2 u}{\partial t^2} = C^2 \frac{\partial^2 u}{\partial x^2} \tag{2.78}$$

which is perturbed by

$$\frac{1}{\lambda} \frac{\partial u}{\partial t} \tag{2.79}$$

where $\partial u / \partial x$ is the longitudinal or shear strain and $C = \sqrt{\sigma/\rho}$ is a shear wave or sound speed. For incompressible fluids, the sound speeds are infinite.

By declaration, fluids which obey a constitutive equation of the type

$$\lambda \frac{D\tau}{Dt} + \tau = 2\mu D[x] \tag{2.80}$$

may be described by Maxwell's models. These models are not unique; they differ in that various invariant derivatives $D\tau/Dt$ can be defined. Up to now, the invariance was not considered. All the invariant nonlinear derivatives reduce to partial time derivatives when linearized at states of rest [66]:

$$\frac{D\tau}{Dt} \to \frac{\partial \tau}{\partial t} \tag{2.81}$$

2.1.4.3 The Voigt or Kelvin Model

This model consists of the spring and dashpot connected in parallel. Figure 2.19b shows the circuit. Hence, the force is given by

$$\sigma_1 + \sigma_2 = \sigma \tag{2.82}$$

where the elastic element is $G\gamma$, and the force in the viscous element is $\mu_2 \partial\gamma/\partial t$; and when in parallel, it follows

$$\sigma = G\gamma + \mu_2 \frac{\partial\gamma}{\partial t} \tag{2.83}$$

This element is instantaneously viscous because, by design, the deformation $\partial\gamma/\partial t$ of the spring and dashpot occurs simultaneously. The dashpot must work in each and every deformation. If a constant force is applied, the deformation will be damped by viscosity and the system will come to equilibrium with $\sigma = G\gamma$, as in an elastic body. The Voigt model is "good" for viscoelastic solids and not for liquids.

2.1.4.4 Jeffrey's Model

This model is presented by the dashpot and Voigt model connected in series. Figure 2.19c shows this circuit. For this design, the deformation is given by

$$\frac{\partial\gamma}{\partial t} = \frac{\partial\gamma_1}{\partial t} + \frac{\partial\gamma_2}{\partial t} \tag{2.84}$$

The forces in both elements are the same:

$$\sigma = \mu_1 \frac{\partial\gamma_1}{\partial t} = G\gamma_2 + \mu_2 \frac{\partial\gamma_2}{\partial t} \tag{2.85}$$

Equations (2.84) and (2.85) are the equations for σ, γ, γ_1, and γ_2, and after eliminating γ_1 and γ_2, it follows that

$$\frac{\mu_1 + \mu_2}{G} \frac{\partial\sigma}{\partial t} + \sigma = \mu_1 \left(\frac{\partial\gamma}{\partial t} + \frac{\mu_2}{G} \frac{\partial^2\gamma}{\partial t^2} \right) \tag{2.86}$$

The Jeffrey's model is viscous because purely elastic deformation of the element is impossible by design, as in the Voigt model. Viscosity is active in every deformation. Jeffrey's element cannot sustain a constant force in equilibrium; the force must relax; therefore, Jeffrey's model is "good" for fluids and not for solids.

If a relaxation time λ_1 is given by

$$\lambda_1 = \frac{\mu_1 + \mu_2}{G} \tag{2.87}$$

and a retardation time is written as

$$\lambda_2 = \frac{\mu_2}{G} \tag{2.88}$$

then for $\lambda_2 = 0$ ($\mu_2 = 0$), Maxwell's model arises from Jeffrey's. Further on, when the relaxation and retardation times are equal,

$$\lambda_1 = \lambda_2 \tag{2.89}$$

then Newtonian fluids arise from Jeffrey's model.

2.1.5 Electroviscosity and Electroviscoelasticity of Liquid-Liquid Interfaces

Electroviscosity and *electroviscoelasticity* are terms that may be broadly defined as dealing with fluid flow effects on physical, chemical, and biochemical processes. The hydrodynamic and/or electrodynamic motion is considered in the presence of both potential fields (elastic forces) and nonpotential fields (resistance forces). The elastic forces are gravitational, buoyancy, and electrostatic-electrodynamic (Lorentz), and the resistance forces are continuum resistance-viscosity, and electrical resistance-impedance.

According to the classical deterministic approach, the phases that constitute the multiphase dispersed systems are assumed to be a continuum, that is, without discontinuities inside the one entire phase, homogeneous, and isotropic. The principles of conservation of mass, momentum, energy, and charge are used to define the state of a real fluid system quantitatively. In addition to the conservation equations, which are insufficient to define the system uniquely, statements on the material behavior are also required. These statements are termed *constitutive relations*, for example, Fourier's law, Fick's Law, and Ohm's law.

In general, the constitutive equations are defined empirically, although the coefficients in these equations (e.g., viscosity coefficient, heat conduction coefficient, and complex resistance coefficient/impedance) may be determined at the molecular level. Often, these coefficients are determined empirically from related phenomena; therefore, such a description of the fluid state is termed a *phenomenological description* or *phenomenological model*.

Sometimes, particular modifications are needed when dealing with a finely dispersed systems. An example is Einstein's modification of the Newtonian viscosity coefficient in dilute colloidal suspensions [66]. Further on is Smoluchowski's modification of Einstein's relation for particles carrying electric double layers, [47], and a recent, more profound elaboration of the entropic effects [11].

3

Particular

3.1 Theory of Electroviscoelasticity

3.1.1 Previous Work

In normal viscous fluids, only the rate of deformation is of interest. In the absence of external and body forces, no stresses are developed and there is no means of distinguishing between a natural state and deformed state [65]. It is rather disturbing to think of the very large overall deformations obtained in the flow of fluids being associated in any way with substances that have elasticity. The rationalization lies in realizing that for the substances considered here, the behavior is essentially that of a fluid; although much translation and rotation may occur, the "elastic" distortion of the elemental volumes around any point is generally small. This "elastic" distortion or material's strain is nevertheless present and is a feature that cannot be neglected. It is responsible for the recovery of reverse flow after the removal of applied forces and for all the other non-Newtonian effects [68]. These distortions or strains are determined by the stress history of the fluid and cannot be specified kinematically in terms of the large overall movement of the fluid. Another way of looking at the situation is to say that the natural state of the fluid changes constantly in flow and tries to catch up with the instantaneous state or the deformed state. It never quite succeeds in doing so, and the lag is a measure of the memory or the elasticity. In elastic solids, the natural state does not change and there is perfect memory [65].

The entropy of elasticity of a droplet is a measure of the increase in the available volume in configuration space. This increase occurs with a transition from a rigid, regular structure to an ensemble of states that include many different structures. If the potential wells in the liquid state were as narrow as those in the solid state, and if each of those potential wells were equally populated and corresponded to a stable amorphous structure (and vice versa), then the entropy of elasticity would be a direct measure of the increase in number of wells, or a direct measure of the number of available structures [7,48].

3.1.2 Structure: Electrified Interfaces—a New Constitutive Model of Liquids

3.1.2.1 Classical Approach to the Existence of Multiphase Dispersed Systems

Following a classical deterministic approach, the phases that constitute a multiphase dispersed system are assumed to be a continuum, that is, without discontinuities inside the entire phase, which is considered homogeneous and isotropic [11,36–42]. Therefore, the basic laws, for example, the conservation of mass, Cauchy's first and second laws of motion, and the first and second laws of thermodynamics, are applicable.

According to the classical approach, the behavior of liquid-liquid interfaces in fine dispersed systems is based on an interrelation between three forms of "instabilities." These are sedimentation, flocculation or coagulation, and coalescence. These events, represented schematically in Figure 3.1, can be understood as a kind of interaction between the liquid phases involved [11,36–45,47–52].

Furthermore, the forces responsible for sedimentation and flocculation are gravity and van der Waals forces of attraction, respectively, and the forces responsible for coalescence are not well known [48–52], although some suggestions have been made recently [38,48].

3.1.2.2 A New Approach to the Existence of Finely Dispersed Systems

A new, deterministic approach discusses the behavior of liquid-liquid interfaces in fine dispersed systems as an interrelation between three other forms of "instabilities." These are rigid, elastic, and plastic [38, 48–52]. Figure 3.2 shows the events that are understood as interactions between the internal/immanent and the external/incidental periodical physical fields.

Since both electric/electromagnetic and mechanical physical fields are present in a droplet, they are considered as immanent or internal, whereas ultrasonic, temperature, or any other applied periodical physical fields are considered as incidental or external. Hereafter, the rigid form of instability comprises the possibility of two-way disturbance spreading, or dynamic

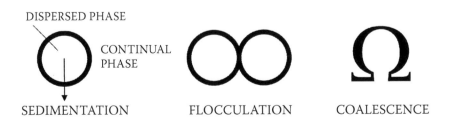

FIGURE 3.1
The classical approach, "instabilities," sedimentation, flocculation, and coalescence.

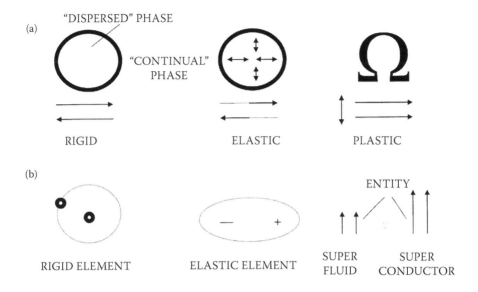

FIGURE 3.2

A new approach: (a) "instabilities," rigid, elastic, and plastic; and (b) the constructive elements of phases.

Source: Reprinted from Spasic, A. M., Jokanovic, V., Krstic, D. N., A theory of electroviscoelasticity: A new approach for quantifying the behavior of liquid-liquid interfaces under applied fields, *J. Colloid Interface Sci.*, 186, 1997, p. 435, with permission from Academic Press.

equilibrium. This form of instability, when all forces involved are in equilibrium, permits a two-way disturbance spreading (propagation or transfer) of entities either by tunneling (low energy dissipation and occurrence probability) or by induction (medium or high energy dissipation and occurrence probability). A classical particle or system could not penetrate a region in which its energy would be negative, that is, barrier regions in which the potential energy is greater than the system energy. In the real world, however, a wave function of significant amplitude may extend into and beyond such a region. If the wave function extends into another region of positive energy, then the barrier is crossed with some probability; this process is termed *tunneling* (since the barrier is penetrated rather than climbed). The elastic form of instability comprises the possibility of reversible disturbance spreading, with or without hysteresis. Finally, the plastic form of instability comprises the possibility of irreversible disturbance spreading with a low or high intensity of influence between two entities. *Entity* is the smallest indivisible element of matter that is related to the particular transfer phenomenon. The entity can be a differential element of mass, or demon; or a phonon as quanta of acoustic energy; or an infon as quanta of information; or a photon; or an electron.

Now, a disperse system consists of two phases, "continuous" and "dispersed." The continuous phase is modeled as an infinitely large number of harmonic electromechanical oscillators with low strength interactions among them. Furthermore, the dispersed phase is a macrocollective consisting of a finite number of microcollectives, i.e., harmonic electromechanical oscillators (clusters) with strong interactions between them. The *cluster* can be defined as the smallest repetitive unit that has a character of integrity. Clusters appear in micro- and nanodispersed systems. The microcollective consists of the following elements: rigid elements (atoms or molecules), elastic elements (dipoles or ions that may be recombined), and entities (as the smallest elements) [48,49,51].

Validation of these theoretical predictions will be corroborated experimentally by means of electrical interfacial potential (EIP) measurements and nuclear magnetic resonance (NMR) spectroscopy [11,38,48–52].

3.1.2.2.1 Model Development

The secondary liquid-liquid droplet or droplet-film structure is considered as a macroscopic system with its internal structure determined by the way the molecules (ions) are tuned (structured) into the primary components of a cluster configuration. How the tuning or structuring occurs depends on the involved physical fields, potential (elastic forces) and nonpotential (resistance forces). All these microelements of the primary structure can be considered as electromechanical oscillators assembled into groups, so that excitation by an external physical field may cause oscillations at the resonant/characteristic frequency of the system itself (coupling at the characteristic frequency) [38,48,49].

Figure 3.3 shows a series of graphical sequences to facilitate the understanding of the proposed structural model of electroviscoelastic liquids. The electrical analog (Figure 3.3a) consists of passive elements (R, L, and C) and an active element (the emitter-coupled oscillator W). Further on, the emitter-coupled oscillator is represented by the equivalent circuit, as shown in Figure 3.3b. Figure 3.3c shows the electrical (oscillators j) and mechanical (structural volumes V_j) analogs when they are coupled between them, for example, in the droplet. Now, the droplet consists of a finite number of structural volumes or spaces/electromechanical oscillators (clusters) V_j, a finite number of excluded surface volumes or interspaces V_s, and a finite number of excluded bulk volumes or interspaces V_b. Furthermore, the interoscillator or intercluster distance or internal separation S_i represents the equilibrium of all forces involved (electrostatic, solvation, van der Waals, and steric [6]). The external separation S_e is introduced as a permitted distance when the droplet is in interaction with any external periodical physical field. The rigidity droplet boundary R presents a form of droplet instability when all forces involved are in equilibrium. Nevertheless, two-way disturbance spreading (propagation or transfer) of entities occurs, either by tunneling (low energy dissipation)

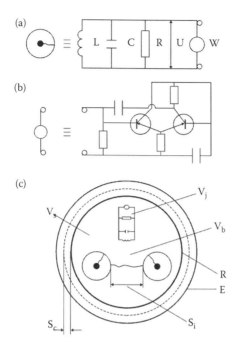

FIGURE 3.3

Graphical interpretation of the structural model.

Note: (a) Electrical and mechanical analog of the microcollective/cluster; (b) equivalent circuit for the emitter coupled oscillator; and (c) the macrocollective: a schematic cross section of the droplet and its characteristics (V_j: structural volumes/clusters; V_s: excluded surface volumes/interspaces; V_b: excluded bulk volumes/interspaces; S_i: internal separation; S_e: external separation; R: rigidity droplet boundary; and E: elasticity droplet boundary).

or by induction (medium or high energy dissipation). The elasticity droplet boundary E represents a form of droplet instability when the equilibrium of all forces involved is disturbed by the action of any external periodical physical field, but the droplet still exists as a dispersed phase. In the region between the rigidity and elasticity droplet boundaries, a reversible disturbance spreading occurs. After the elasticity droplet boundary, the plasticity as a form of droplet instability takes place; then the electromechanical oscillators or clusters do not exist any more and the beams of entities or attoclusters appear. In this region, one-way propagation of entities occurs.

Considering all presented arguments and comments, the probability density function (PDF) in general form can be expressed by

$$F_d(V) = F_d(V_j) + [F_d(V_s) + F_d(V_b)] \tag{3.1}$$

where the first term on the right-hand side of the equation is due to energy effects, and the second term (consisting of two subterms) is due to entropic effects; index j is related to the structural volumes or energies, and indexes s and b are related to the excluded surface and bulk volumes or energies.

An alternative expression of this PDF, considering Figure 3.1c, may be written as

$$\Delta_{1,2}(V) = \sum_{i=3}^{n}(V_j)_i + \Delta_{1,2}\left[\sum_{i=1}^{n-3}(V_b)_i + \sum_{i=3}^{n}(V_s)_i\right] \tag{3.2}$$

Therefore, the number of clusters V_j remains constant while the droplet passes through the rigid and elastic form of "instabilities." Differences in volumes or energies V_b and V_s can occur only in the entropic part, that is, the internal separation S_i (Figure 3.1c) changes (increases or decreases) during the transition of the droplet from rigid to elastic or vice versa. Consequently, the external separation S_e decreases or increases depending on the direction of transition.

3.1.2.3 Classical Assumptions for Interfacial Tension Structure and for Partition Function

The droplet is considered as a unique thermodynamic system that can be described by a characteristic free energy function expressed by

$$\Delta G = (\sigma_i + T\Delta S) = \left[\sigma_i - T\left(\frac{d\sigma_i}{dT}\right)_{\chi_i}\right] = -kT \ln Z_p \tag{3.3}$$

According to quantum mechanical principles, the droplet possesses vacancies or "free volumes," and the relation for interfacial tension can be written as

$$\sigma_i = \frac{G^s - G^b}{\kappa^0} = \frac{1}{\kappa^0}\left[(\Phi^s - \Phi^b) + N^s kT \ln\left(\frac{V_f^b}{V_f^s}\right)\right] \tag{3.4}$$

where Φ^s and Φ^b are the overall energies of N heavy-phase molecules in their ideal positions, s on the surface and b in the bulk. N^s is the number of molecules on the surface, V_f^s and V_f^b are "free volumes" of the molecules on the surface and in the bulk, and κ^0 is the surface of "free surface" [19,20,38,48,66–86].

The phenomenological meaning of the given interfacial tension structure is in agreement with the "free volume" fluid model. Hence, a fluid is a system with ideal or ordered neighborhood elements distribution and with discontinuities of the package density (boundaries of the subsystems or microcollectives as some particular physical systems) [38,48].

3.1.2.4 Postulated Assumptions for an Electrical Analog

Here, postulated assumptions for an electrical analog are summarized as follows:

1. The droplet is a macrosystem (collective of particles) consisting of structural elements that may be considered as electromechanical oscillators.

2. Droplets as microcollectives undergo tuning or coupling processes, and so build the droplet as a macrocollective.

3. The external physical fields (temperature, ultrasonic, electromagnetic, or any other periodic) cause the excitation of a macrosystem through the excitation of microsystems at the resonant/characteristic frequency, where elastic and/or plastic deformations may occur.

Hence, the study of the electromechanical oscillators is based on electromechanical and electrodynamic principles. At first, during the droplet formation it is possible that the serial analog circuits are more probable, but later, as a consequence of tuning and coupling processes the parallel circuitry becomes dominant. Also, since the transfer of entities by tunneling (although with low energy dissipation) is much less probable, it is sensible to consider the transfer of entities by induction (medium or high energy dissipation).

A nonlinear integral-differential equation of the van der Pol type represents the initial electromagnetic oscillation

$$C\frac{dU}{dt} + \left(\frac{U}{R} - \alpha U\right) + \gamma U^3 + \frac{1}{L}\int U\, dt = 0 \tag{3.5}$$

where U is the overall potential difference at the junction point of the spherical capacitor C and the plate, L is the inductance caused by potential difference, and R is the Ohm resistance (resistance of the energy transformation, electromagnetic into the mechanical or damping resistance), t is the time, and α and γ are constants determining the linear and nonlinear parts of the characteristic current and potential curves. U_0, the primary steady-state solution of this equation, is a sinusoid of frequency close to $\omega_0=1/(LC)^{0.5}$ and amplitude $A_0=[(\alpha-1/R)/3\gamma/4]^{0.5}$.

The noise in this system, due to linear amplification of the source noise (the electromagnetic force assumed to be the incident external force, which initiates the mechanical disturbance), causes the oscillations of the "continuum" particle (the molecule surrounding the droplet or droplet-film structure), which can be represented by the particular integral

$$C\frac{dU}{dt} + \left(\frac{1}{R} - \alpha\right)U + \gamma U^3 + \frac{1}{L}\int U\, dt = -2A_n \cos\omega t \tag{3.6}$$

where ω is the frequency of the incident oscillations.

Finally, considering the droplet or droplet-film structure formation, "breathing," and/or destruction processes, and taking into account all the noise frequency components, which are included in the driving force, the corresponding equation is given by

$$C\frac{dU}{dt} + \left(\frac{1}{R} - \alpha\right)U + \frac{1}{L}\int U\,dt + \gamma U^3 = i(t) = \frac{1}{2\pi}\int_{-\infty}^{\infty} \exp(j\omega t)A_n(\omega)\,d\omega \qquad (3.7)$$

where $i(t)$ is the noise current and $A_n(\omega)$ is the spectral distribution of the noise current as a function of frequency [1,2].

In the case of nonlinear oscillators, however, the problem of determining the noise output is complicated by the fact that the output is fed back into the system, thus modifying in a complicated manner the effective noise input [1,2]. The noise output appears as an induced anisotropic effect.

3.1.3 Dynamics: Physical Formalism

A number of theories that describe the behavior of liquid-liquid interfaces have been developed and applied to various dispersed systems, for example, those of Stokes, Reiner-Rivelin, Ericksen, Einstein, Smoluchowski, and Kinch [21,22,38,47,48,83,84,86,89,92,93,95,98–109].

The reader is suggested to review some of the following topics for a better understanding of the present theory: potential energy surfaces (PES; Dirac, Millikan, Feynman, Schwinger, Tomonaga, Born-Oppenheimer, etc.), quantum electrodynamics (QED; Schrodinger), molecular mechanics (MM2-MM3-CSC; Allinger, Lii, and Cambridge Science Computation), and transient state theories (TST; Wigner, etc.) [7,23–26,68,87,102–109].

Figure 3.4 shows a series of graphical sequences that are supposed to facilitate the understanding of the proposed theory of electroviscoelasticity. This theory describes the behavior of electrified liquid-liquid interfaces in finely dispersed systems, and is based on a new constitutive model of liquids [38, 48–52]. If an incident periodical physical field (Figure 3.4b), for example, electromagnetic, is applied to the rigid droplet of Figure 3.4a, then the resultant, equivalent electrical circuit can be represented as shown in Figure 3.4c. The equivalent electrical circuit, rearranged under the influence of an applied physical field, is considered as a parallel resonant circuit coupled to another circuit, such as an antenna output circuit. Now again, the initial electromagnetic oscillation is represented by the differential equations, Eqs. (3.5) and (3.6), and when the nonlinear terms are omitted and/or superposed, the simpler linear equation is given by

$$C\frac{dU}{dt} + \left(\frac{1}{R} - \alpha\right)U + \frac{1}{L}\int U\,dt+ = -2A_n\cos\omega t \qquad (3.8)$$

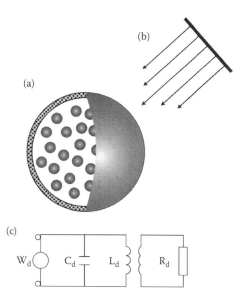

FIGURE 3.4

Definition sketch for understanding the theory of electroviscoelasticity.

Note: (a) Rigid droplet; (b) incident physical field, e.g., electromagnetic; and (c) equivalent electrical circuit-antenna output circuit. W_d represents the emitter-coupled oscillator, C_d, L_d, and R_d are capacitive, inductive, and resistive elements of the equivalent electrical circuit, respectively. Subscript d is related to the particular diameter of the droplet under consideration.

Source: Spasic, A. M., Lazarevic, M. P., Krstic, D. N., Theory of electroviscoelasticity, in: *Finely Dispersed Particles: Micro-, Nano, and Atto-Engineering*, A. M. Spasic and J. P. Hsu, Eds., CRC Press/Taylor & Francis, Boca Raton, FL, 2006, 1–23 and 371–393; and Spasic, A. M., Electroviscoelasticity of liquid-liquid interfaces, in: *Interfacial Electrokinetics and Electrophoresis*, A. V. Delgado, Ed., Marcel Dekker, New York, 2002, 837–867. Courtesy of Marcel Dekker, Inc.

with a particular solution resulting in the following expression for the amplitude:

$$A = \frac{2\omega C A_n}{\left[4(\omega_0 - \omega)^2 + \left(\frac{1}{R} - \alpha\right)^2 \right]^{0.5}} \tag{3.9}$$

And for all the noise frequency components, the simpler linear equation is given by

$$C\frac{dU}{dt} + \left(\frac{1}{R} - \alpha\right)U + \frac{1}{L}\int U\, dt = i(t) = \frac{1}{2\pi}\int_{-\infty}^{\infty} \exp(j\omega t) A_n(\omega)\, d\omega \tag{3.10}$$

with the particular solution expressed by

$$U_n = \frac{i\omega A_n \exp(i\omega t)}{C(\omega_0^2 - \omega^2) + i(\frac{1}{R} - \alpha)\omega} - \frac{i\omega A_n \exp(-i\omega t)}{C(\omega_0^2 - \omega^2) + i(\frac{1}{R} - \alpha)\omega} \qquad (3.11)$$

Again, looking at Eqs. (3.5)–(3.11) and Figure 3.4, after the cluster's rearrangement the resultant equivalent electrical circuit can be represented as shown in Figure 3.4c. Figure 3.5 shows the behavior of the circuit depicted in Figure 3.4c, using the correlation impedance-frequency-arbitrary droplet diameter.

Since all events occur at the resonant/characteristic frequency, depending on the amount of coupling, the shape of the impedance-frequency curve is judged using the factor of merit or Q factor [82]. The Q factor primarily determines the sharpness of resonance of a tuned circuit, and may be represented as the ratio of the reactance to the resistance, as follows:

$$Q = \frac{2\pi f L}{R} = \frac{\omega L}{R} \qquad (3.12)$$

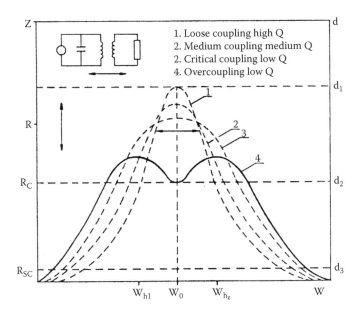

FIGURE 3.5
Impedance of the equivalent electric circuit versus its frequency.

Source: From Spasic, A. M., Jokanovic, V., Krstic, D. N., A theory of electroviscoelasticity: A new approach for quantifying the behavior of liquid-liquid interfaces under applied fields, *J. Colloid Interface Sci.*, 186, 434–446, 1997, with permission from Academic Press.

Furthermore, the impedance Z can be related to the factor of merit as it is given by the following equations:

$$Z = \frac{(2\pi f R)^2}{R} = \frac{(\omega L)^2}{R} \tag{3.13}$$

and

$$Z = \omega L Q \tag{3.14}$$

From these expressions and Figure 3.5, it can be seen that the impedance of a circuit is directly proportional to its effective Q at resonance. Also at the resonant frequency ω_0 the impedance Z is equal to the resistance R, R_c/ critical and R_{sc}/supercritical, respectively. These resistances and Z-ω curves correspond to the various levels of coupling: (1) loose coupling and high Q, (2) medium coupling and medium Q, (3) critical coupling and low Q, and (4) overcoupling and low Q. ω_{h1} and ω_{h2} represent the hump frequencies that appear during overcoupling (curve 4). On the right axes of Figure 3.5, the corresponding critical diameters d_1, d_2, and d_3 are arbitrary plotted.

Further on, the electrical energy density w_e inside the capacitor is given by

$$w_e = \frac{1}{8\pi} \varepsilon E^2 \tag{3.15}$$

where ε is the dielectric constant and E is the electric field, and the magnetic energy density w_m inside the capacitor is given by

$$w_m = \frac{1}{8\pi} \mu_e H^2 \tag{3.16}$$

where μ_e is the magnetic permeability constant and H is the magnetic field; hence, the overall mean energy may be written as

$$\overline{w} = \frac{1}{8\pi} (\varepsilon E^2 + \mu_e H^2) \tag{3.17}$$

The electromagnetic oscillation causes the tuning or structuring of the molecules (ions) in the "electric double layers." The structuring is realized by complex motions over the various degrees of freedom, whose energy contributions depend on the positions of the individual molecules in and around the stopped droplet-film structure under the action of some periodical physical field.

The hydrodynamic motion is considered to be the motion in the potential field (elastic forces) and nonpotential field (resistance forces). There are several possible approaches to correlate the electromagnetic and mechanical oscillations, for example, this motion may be represented by the differential equation

$$\frac{d^2\xi}{dt^2} + 2\beta\frac{d\xi}{dt} - \omega_0^2 = 0 \tag{3.18}$$

where from the relation expressed by

$$m\tilde{a} = k_N\tilde{\xi}_y - r\tilde{v} \tag{3.19}$$

the relations

$$\omega_0^2 = \frac{k_N}{m} \tag{3.20}$$

and

$$2\beta = \frac{r}{m} \tag{3.21}$$

where $k_N\tilde{\xi}_y$ is the potential force (electrostatic/electromagnetic) and $r\tilde{v}$ is the nonpotential force (continuum resistance/viscosity), are derived.

The incident hybrid mechanical oscillation of the ordered group of molecules may be written as

$$\xi = \xi_0\exp(-\beta t)\cos(\omega t + \zeta) \tag{3.22}$$

where the amplitude as a function of time is given by

$$\xi(t) = \xi_0\exp(-\beta t) = \xi_0\exp\left(\frac{-r}{2m}\right)t \tag{3.23}$$

and ξ_0 is initial amplitude. The continuum resistance constant r is given by

$$r = 6\pi r'\mu \tag{3.24}$$

Now, if the electromagnetic force is assumed to be the incident (external) force which initiates mechanical disturbance, then the oscillation of the "continuum" particles (molecules surrounding the droplet-film structure) is described by the differential Eq. (3.18), where ω is the frequency of the incident oscillations. After a certain time, the oscillations of the free oscillators (molecules surrounding the droplet-film structure) tune with the incident oscillator frequency. This process of tuning can be expressed by

$$\Omega = \Omega_{ref} = (\omega_0^2 - 2\beta^2)^{0.5} \tag{3.25}$$

Thereafter, if a wave with high amplitude appears, then the rupture of the droplet-film structure occurs [11].

Further on, the energy of the mechanical wave is expressed by

$$W = W_k + W_p = \frac{1}{2}\rho\Delta V \xi_0^2 \omega^2 \sin^2 \omega\left(t - \frac{\xi}{v}\right) + \frac{1}{2}E_y\Delta V \xi_0^2 \left(\frac{m}{v}\right)\sin^2 \omega\left(t - \frac{\xi}{v}\right) \tag{3.26}$$

where ξ is the mechanical wave function, $1/2\,\rho\xi_0^2\omega^2$ is the density of the mechanical energy, E_y is the modulus of elasticity, and ΔV is the differential element of volume.

During the interaction of the droplet or droplet-film structure with an incident periodical physical field at the instant of equilibrium, the mean energy can be written as

$$\overline{w} = \frac{1}{8\pi}(\varepsilon E^2 + \mu H^2) = \frac{1}{2}\rho\xi_0^2\omega^2 \tag{3.27}$$

and the frequency of the incident wave can be expressed by

$$\omega = \left[\frac{1}{4\pi\rho\xi^2}(\varepsilon E^2 + \mu H^2)\right]^{0.5} \tag{3.28}$$

Furthermore, the equivalency of the adequate mean energies, at the instant of equilibrium, can be related to a characteristic function of free energy expressed by

$$\Delta G = \overline{w} = -kT \ln Z_p \tag{3.29}$$

where the partition function for this oscillator is derived from

$$W = hf\,\frac{\partial \ln Z_p}{\partial \theta} \tag{3.30}$$

and

$$Z_p = \frac{Q_N}{\lambda^{3N}} = \sum_{j=0}^{\infty}\exp\left[-\left(j+\frac{1}{2}\right)\Theta\right] = \frac{1}{2}\cosh\frac{\Theta}{2} \tag{3.31}$$

and j is a number of the identical oscillators, where each is given by

$$\Theta = \frac{hf}{kT} \tag{3.32}$$

λ is a free path between two collisions expressed by

$$\lambda^2 = \frac{h^2}{2\pi mkT} \tag{3.33}$$

and Q_N is a configuration integral. Considering the assumption of the determined geometrical structure of the cluster type one may derive the internal structure of the droplet as a number of microcollectives associated in the macrocollective. The droplet consists of N elementary units, structure elements, or molecules, which constitute the macrocollective of l_1, l_2, l_3, ..., l_N constitutive elements/clusters. The configuration integral expressing the energy distribution inside the system may be written as

$$Q_N = \frac{1}{N!} \int W_N(r^N) \; dr^N \tag{3.34}$$

where the Boltzmann probability factor W_N may be written as

$$W_N(r^N) = \sum \prod U_l(r^N) \tag{3.35}$$

Further,

$$Q_N = \sum \prod_{l=1}^{N} \frac{(Ubl)^{ml}}{ml!} \tag{3.36}$$

and

$$\sum lml = N \tag{3.37}$$

where ml is the number of groups consisting of $(1-l)$ elements, and bl is the cluster integral, which, in general form for N elements, may be expressed by

$$bl = (bl!)^{-1} \int U_l(r_1, r_2, r_3, \ldots, r_l) \; dr_1, \ldots, dr_l \tag{3.38}$$

Now, the Boltzmann probability factor may be written as

$$W_N(r^N) = \frac{C}{2}(N - \Omega)\zeta(a) \tag{3.39}$$

Finally, the configuration integral may be expressed by

$$Q_n = \sum \int_{l_1} dr_1 \int_{l_2} dr_2 \cdots \int_{l_N} dr_N \exp\left[\frac{-\Phi(r^N)}{kT}\right] \tag{3.40}$$

The presented theory has been applied to the representative experimental system described in Section 2.1.1.1 and 4.1.1.1, "Description of the Physical-Chemical System." Validation of the theoretical predictions was corroborated experimentally by means of EIP measurements, and by means of NMR spectroscopy. These methods and apparatus are briefly presented in Chapter 4, Sections 4.1 and 4.2. The obtained experimental results were in fair agreement with the postulated theory. Measured, calculated, and estimated data are presented in Chapter 4, Section 4.3 [1,11,38,48].

3.1.4 Mathematical Formalisms

3.1.4.1 The Stretching Tensor Model

Now, using the presented propositions and electromechanical analogies, an approach to non-Newtonian behaviors and to electroviscoelasticity is to be introduced. When Eq. (2.7) is applied to the droplet when it is stopped, for example, as a result of an interaction with some periodical physical field, the term on the left-hand side becomes equal to zero.

$$\rho \frac{D\tilde{u}}{Dt} = \sum_i \tilde{F}_i(dxdydz) + d\tilde{F}_s \tag{3.41}$$

Furthermore, if the droplet is in the state of "forced" levitation, and the volume forces balance each other, then the volume force term is also equal to zero [1,11,38,48,49,62]. It is assumed that the surface forces are, for the general case that includes the electroviscoelastic fluids, composed of interaction terms expressed by

$$d\tilde{F}_s = \tilde{T}^{ij}d\tilde{A} \tag{3.42}$$

where the tensor T^{ij} is given by

$$T^{ij} = -\alpha_0\delta^{ij} + \alpha_1\delta^{ij} + \alpha_2\zeta^{ij} + \alpha_3\zeta_k^i\zeta^{kj} \tag{3.43}$$

where T^{ij} is composed of four tensors, δ^{ij} is the Kronecker symbol, ζ^{ij} is the stretching tensor, and $\zeta^i_k \zeta^{kj}$ is the stretching coupling tensor. In the first isotropic tensor, the potentiostatic pressure $\alpha_0=\alpha_0(\rho, U)$ is dominant and the contribution of the other elements is neglected. Here U represents hydrostatic or electrostatic potential. In the second isotropic tensor, the resistance $\alpha_1=\alpha_1 (\rho, U)$ is dominant and the contribution of the other elements is neglected. In the third stretching tensor, its normal elements $\alpha_2\sigma$ are due to the interfacial tensions and the tangential elements $\alpha_2\tau$ are presumed to be of the same origin as the dominant physical field involved. In the fourth stretching coupling tensor, there are normal, $\alpha_3\sigma^i_k$, and $\alpha_3\sigma^{kj}$ elements, and tangential $\alpha_3\tau^i_k$ and $\alpha_3\tau^{kj}$ elements, that are attributed to the first two dominant periodical physical fields involved. Now, the general equilibrium condition for the dispersed system with two periodical physical fields involved may be derived from Eq. (3.50), and may be expressed by

$$\tau_d = \frac{-\alpha_0 + \alpha_1 + \alpha_2\left(\frac{\sigma}{d}\right) + \alpha_3\left(\frac{\sigma}{d}\right)}{2(\alpha_2 + \alpha_3)} \tag{3.44}$$

where τ_d are the tangential elements of the same origin as those of the dominant periodical physical field involved. Figure 3.6 shows the schematic equilibrium of surface forces at any point of a stopped droplet-film structure while

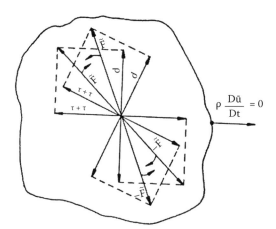

FIGURE 3.6

Balance of surface forces at any point of a stopped droplet-film structure while in interaction with some periodic physical field (a two-dimensional projection).

Note: F: the projection of the resultant surface forces, a vector in three N-dimensional configuration space; τ: the tangential components; and σ: the normal components.

Source: From Spasic, A. M., Jokanovic, V., Krstic, D. N., A theory of electroviscoelasticity: A new approach for quantifying the behavior of liquid-liquid interfaces under applied fields, *J. Colloid Interface Sci.*, 186, 434–446, 1997, with permission from Academic Press.

in interaction with some periodical physical field [1,11,62]. Note that for dispersed systems consisting of, or behaving as, Newtonian fluids, $\alpha_3=\alpha_3(\rho,U)$ is equal to zero.

The processes of formation and destruction of the droplet or droplet-film structure are nonlinear. Therefore, the viscosity coefficients μ_i ($i = 0,1,2$), where each consists of bulk, shear, and tensile components, when correlated to the tangential tensions of mechanical origin τ_v, can be written as

$$\tau_v = \mu_0 \frac{du}{dx} + \mu_1 \frac{d^2u}{dx^2} + \mu_2 \left(\frac{du}{dx}\right)^2 \tag{3.45}$$

where u is the velocity, and x is one of the space coordinates.

Using the electrical analog, the impedance coefficients Z_i ($i = 0,1,2$), where each consists of Ohmic, capacitive, and inductive components, will be correlated with the tangential tensions of electrical origin τ_e, as follows:

$$\tau_e = Z_0 \frac{d\phi_e}{dt} + Z_1 \frac{d^2\phi_e}{dt^2} + Z_2 \left(\frac{d\phi_e}{dt}\right)^2 \tag{3.46}$$

where ϕ_e is the electron flux density, and t is the time coordinate.

More detailed discussion about derivation of these equations can be found in References [1,11,38,48,62].

3.1.4.2 The van der Pol Derivative Model: The Fractional Approach

Fractional derivatives provide an excellent instrument for the description of memory and hereditary properties of various materials and processes [110–126]. This is the main advantage of fractional derivatives compared to the classical integer-order models, in which such effects are in fact neglected. The mathematical modeling and simulation of systems and processes, based on the description of their properties in terms of fractional derivatives, naturally lead to differential equations of fractional order and to the necessity to solve such equations.

3.1.4.2.1 Fundamentals of Fractional Calculus

The idea of fractional calculus has been known since the development of regular calculus, with the first reference probably being associated with correspondence between Gottfried Leibniz and the Marquis de L'Hôpital in 1695, where a differentiation of order one-half was discussed. Along the years, great mathematicians such as Leonhard Euler, Joseph Fourier, Niels Abel, and others did some work on fractional calculus that, surprisingly, remained as a sort of curiosity. Further on, the theory of fractional-order derivatives was developed mainly in the nineteenth century. In his seven-hundred-page book on calculus, published in 1819, Silvestre Lacroix developed the formula for the nth derivative of $y = x^m$, where m is a positive integer, and $D^n x^m = (m!/(m-n)!)x^{m-n}$, where $n(\leq m)$ is an integer.

The modern epoch started in 1974, when a consistent formalism of fractional calculus was developed by K. B. Oldham and J. Spanier [110], in which fractional (noninteger) calculus is a generalization of ordinary differential and integral calculus. Only in the last few decades, finally, scientists and engineers have realized that such fractional differential equations provide a natural framework for the discussion of various kinds of problems modeled by fractional differential and fractional integral equations, that is, they provide more accurate models of systems under considerations. A number of authors have demonstrated applications of fractional calculus in various fields, such as physics, chemistry, and engineering [110,111,127]; also, a few works dealing with the application of this mathematical tool in signal processing, anomalous diffusion [128], and control theory [111] have been published.

Concerning a continuous-time modeling, fractional calculus may be of great interest, for example, in problems related to viscoelasticity [129, 130], electrochemical processes [131,132], polymer chemistry, and heat, mass, momentum, and electron transfer phenomena [119]. The main reason for the success of the theory in these cases is that these new fractional-order models are more accurate than integer-order models, that is, there are more degrees of freedom in the fractional-order model. Furthermore, fractional derivatives provide an excellent instrument for the description of a memory and for hereditary properties of various materials and processes due to the existence of a "memory term" in a model. This memory term insures the history and its impact to the present and future. Fractional-order models have an unlimited memory compared to the integer-order models, which have a limited memory.

The fractional integro-differential operators (fractional calculus) present a generalization of integration and derivation to non-integer-order (fractional) operators. At first, one can generalize the differential and integral operators into one fundamental D_t^p operator t, which is known as fractional calculus:

$$
_aD_t^p = \begin{cases} \dfrac{d^p}{dt^p} & \Re(p) > 0, \\[2mm] 1 & \Re(p) = 0, \\[2mm] \displaystyle\int_\alpha^t (d\tau)^{-p} & \Re(p) < 0. \end{cases} \tag{3.47}
$$

Several definitions of a fractional derivative have been proposed. These definitions include Riemann-Liouville, Grunwald-Letnikov, Weyl, Caputo, Marchaud, and Riesz fractional derivatives [110,111,133]. The two definitions generally used for the fractional differintegral are the Grunwald-Letnikov

(GL) definition and the Riemann-Liouville (RL) definition [110,111]. The original GL definition of fractional derivatives is given by a limit, namely,

$$_aD_t^p f(t) = \lim_{h \to 0} \frac{1}{h^p} \sum_{j=0}^{[(t-a)/h]} (-1)^j \binom{p}{j} f(t - jh)$$ (3.48)

where a and t are the limits of the operator, and $[x]$ means the integer part of x. The integral version of GL is defined by

$$_aD_t^p f(t) = \sum_{k=0}^{n-1} \frac{f^{(k)}(0)t^{-p+k}}{\Gamma(-p+k+1)} + \frac{1}{\Gamma(n-p)} \int_a^t \frac{f^{(n)}(\tau)}{(t-\tau)^{p-n+1}} d\tau,$$ (3.49)

The RL definition of fractional derivatives is given by

$$_aD_t^p f(t) = \frac{1}{\Gamma(n-p)} \frac{d^n}{dt^n} \int_a^t \frac{f(\tau)}{(t-\tau)^{p-n+1}} d\tau,$$ (3.50)

For $(n - 1 < p < n)$ and for the case of $(0 < p < 1)$, the fractional integral is defined as

$$_0D_t^{-p} f(t) = \frac{1}{\Gamma(p)} \int_0^t \frac{f(\tau)}{(t-\tau)^{1-p}} d\tau,$$ (3.51)

where $\Gamma(.)$ is the well-known Euler's gamma function, written as

$$\Gamma(z) = \int_0^\infty e^{-t} t^{z-1} dt, \quad z = x + iy, \quad \Gamma(z+1) = z\Gamma(z)$$ (3.52)

One of the basic properties of the gamma function is that it satisfies the functional equation

$$\Gamma(z+1) = z\Gamma(z), \Rightarrow \Gamma(n+1) = n(n-1)! = n!$$ (3.53)

Also, the chain rule has the form

$$\frac{d^\beta f(g(t))}{dt^\beta} = \sum_{k=0}^\infty \binom{\beta}{k}_\Gamma \left(\frac{d^{\beta-k}}{dt^{\beta-k}} 1\right) \frac{d^k}{dt^k} f(g(t))$$ (3.54)

where $k \in N$ and $\begin{pmatrix} \beta \\ k \end{pmatrix}_\Gamma$ are the coefficients of the generalized binomial

$$\begin{pmatrix} \beta \\ k \end{pmatrix}_\Gamma = \frac{\Gamma(1+\beta)}{\Gamma(1+k)\Gamma(1-k+\beta)} \tag{3.55}$$

Also, Caputo [130] has proposed that one should incorporate the integer-order (classical) derivative of a function x, as they are commonly used in initial-value problems with integer-order equations. In that way, one can use the derivatives of the Caputo type such as

$$\substack{C \\ 0}D_t^p[U(t)] = \frac{d^p U}{dt^p} = \frac{1}{\Gamma(n-p)} \int_0^t \frac{U^{(n)}(\tau)}{(t-\tau)^{n-p-1}} dt, \quad n-1 < p < n, \quad U^{(n)}(\tau) = d^n U/d\tau^n$$

$$\tag{3.56}$$

According to Diethelm et al. [134], Eq. (3.56) had been already introduced by Rabotnov a year before Caputo's paper was published. However, in the literature, the fractional derivative defined by Eq. (3.56) is known as the *Caputo fractional derivative*.

For convenience, *Laplace domain* is usually used to describe the fractional integro-differential operation for solving engineering problems. The formula for the Laplace transform of the RL fractional derivative has the following form:

$$\int_0^\infty e^{-st} {}_0D_t^p f(t)\,dt = s^p F(s) - \sum_{k=0}^{n-1} s^k {}_0D_t^{p-k-1}f(t)\big|_{t=0} \tag{3.57}$$

And the Laplace transform of the Caputo fractional derivative is

$$\int_0^\infty e^{-st} \,{}_0^C D_t^p f(t)\,dt = s^p F(s) - \sum_{k=0}^{n-1} s^{p-k-1} f^{(k)}(0) \tag{3.58}$$

where Eq. (3.57) involves initial conditions $f^{(k)}(0)$ with integer derivatives $f^{(k)}(t)$. In pure mathematics, the RL derivative is more commonly used than the Caputo derivative. In practical applications, the initial conditions ${}_0D_t^{p-k-1}f(t)\big|_{t=0}$ are frequently not available [111], so the Caputo fractional derivative is considered here where one should incorporate derivatives of integer order of the function f as the initial conditions, Eq. (3.57). Recently, Heymans and Podlubny [135] gave some explanations for RL fractional-order initial values, where it is possible to obtain initial values for such initial conditions by appropriate measurements or observations.

From the definition of the RL and Caputo derivatives, one may observe that the relation between two fractional derivatives is

$$\substack{C\\0}D_t^p[U(t)] = {}_0D_t^p[(U - T_{n-1}[U])(t)], \tag{3.59}$$

where $T_{n-1}[U]$ is the Taylor polynomial of order $(n-1)$ for U, centered at 0. So, one can specify the initial conditions in the classical form

$$U^{(k)}(0) = U_0^{(k)}, \quad k = 0, 1, \ldots, n-1 \tag{3.60}$$

The two RL and Caputo formulations coincide when the initial conditions are zero.

For a numerical calculation of fractional-order differintegral operators, one can use the relation derived from the GL definition. This relation has the following form:

$$_{(t-L)}D_t^{\pm p}f(t) \approx h^{\mp p}\sum_{j=0}^{N(t)} b_j^{(\pm p)}f(t-jh) \tag{3.61}$$

where L is the "memory length," h is the step size of the calculation, $N(t) = \min\{[\frac{t}{h}], [\frac{L}{h}]\}$, $[x]$ is the integer part of x, and $b_j^{(\pm p)}$ is the binomial coefficient given by

$$b_0^{(\pm p)} = 1, \quad b_j^{(\pm p)} = \left(1 - \frac{1 \pm p}{j}\right)b_{j-1}^{(\pm p)} \tag{3.62}$$

3.1.4.2.2 Example of Analog Realization of a Fractional Element

Constructing an analog realization of a fractional-order element may be much easier than the discrete ladder circuits proposed so far. If a single material exhibited "fractance" characteristics, then a single component would replace the entire network. In fact, an ideal capacitor does not exist. An ideal dielectric in a capacitor having an impedance of the form $(1/j\omega C)$ would violate causality. It seems sensible to look for dielectric materials exhibiting the more realistic fractional behavior $1/(j\omega C)^\alpha$, where $p \approx 0.5$. Such a component would display "fractance" attributes and could be termed a *fractor*, as opposed to a resistor or capacitor. In this case, a single component would do the job of an entire network of "ideal" components. Moreover, the capacitor's impedance is described by the transfer function

$$Z(s) = 1/Cs^p, \quad 0 < p < 1 \tag{3.63}$$

Also, such circuits can be obtained if generalized models of resistors, capacitors, and induction coils are taken. For example, Schmidt and Drumheller [126] investigated lithium hydrazinium sulfate (LiN2H5SO4) and found that "over a large range of temperature and frequency the real and imaginary parts of the susceptibility are very large (up to $\varepsilon' \approx \varepsilon'' = 10^6$, $\varepsilon = \varepsilon' + j\varepsilon''$) and vary with frequency somewhat as $f^{-1/2}$," and one can get $\varepsilon = \varepsilon_r \sqrt{2}(j\omega)^{-1/2}$. Using the definition of the relationship between the dielectric function and the impedance,

$$Z = \frac{1}{j\omega C_C \varepsilon_r \sqrt{2}(j\omega)^{-1/2}} \tag{3.64}$$

where C_C is the empty cell capacitance. One gets the impedance of the "fractor"

$$Z_F = \frac{K}{s^{1/2}} \tag{3.65}$$

converting to the Laplace notation $j\omega \to s$, where K represents the lumped constants of the impedance.

3.1.4.2.3 Solution of the Representative van der Pol–Linear Model

In an effort to generalize Eq. (3.5), the ordinary time derivative and integral are now replaced with a corresponding fractional-order time derivative and integral of order $p < 1$ [1,2]. Fractional derivatives provide an excellent instrument for the description of memory and hereditary properties of various materials and processes. This is the main advantage of fractional derivatives in comparison with classical integer-order models, in which such effects are in fact neglected. Here, the capacitive and inductive elements, using fractional-order $p \in (0,1)$, enable formation of the fractional differential equation, that is, a more flexible or general model of liquid-liquid interface behavior. Now, a differintegral form using the RL definition is given by

$$_0D_t^p[U(t)] = \frac{d^p U}{dt^p} = \frac{1}{\Gamma(1-p)}\frac{d}{dt}\int_0^t \frac{U(\tau)}{(t-\tau)^p}\,d\tau, \quad _0D_t^{-p}[U(t)] = \frac{1}{\Gamma(p)}\int_0^t \frac{U(\tau)}{(t-\tau)^{1-p}}\,d\tau,$$

$$0 < p < 1 \qquad\qquad\qquad\qquad p > 0$$

$$\tag{3.66}$$

So, in that way one can obtain linear fractional differential equations with zero initial conditions as follows:

$$C_0 D_t^p[U(t)] + \left(\frac{1}{R} - \alpha\right)U + \frac{1}{L}\,_0D_t^{-p}[U(t)] = i(t) \tag{3.67}$$

Applying the Laplace transform to this expression leads to

$$G(s) = \frac{U(s)}{i(s)} = \frac{1}{Cs^p + 1/Ls^{-p} + (1/R - \alpha)} = \frac{s^p}{Cs^{2p} + (1/R - \alpha)s^p + 1/L} \quad (3.68)$$

or

$$G(s) = s^p G_3(s), \quad G_3(s) = \frac{1}{as^{2p} + bs^p + c}, \quad a = C, \ b = (1/R - \alpha), \ c = 1/L \quad (3.69)$$

Further, $G_3(s)$ in the form

$$G_3(s) = \frac{1}{as^{2p} + bs^p + c} = \frac{1}{c} \frac{cs^{-p}}{as^{2p-p} + b} \frac{1}{1 + \frac{cs^{-p}}{as^{2p-p}+b}} = \frac{1}{c} \sum_{k=0}^{\infty} (-1)^k \left(\frac{c}{a}\right)^{k+1} \frac{s^{-pk-p}}{(s^{2p-p} + b/a)^{k+1}}$$

$$(3.70)$$

The term-by-term inversion, based on the general expansion theorem for the Laplace transform [134], and using the two-parameter function of the Mittag-Leffler type, is defined by the series expansion:

$$E_{\alpha,\beta}(z) = \sum_{k=0}^{\infty} \frac{z^k}{\Gamma(\alpha k + \beta)}, \quad (\alpha, \beta > 0). \quad (3.71)$$

The Mittag-Leffler function is a generalization of exponential function e^z, and the exponential function is a particular case of the Mittag-Leffler function. Here is the relationship given in

$$E_{1,1}(z) = \sum_{k=0}^{\infty} \frac{z^k}{\Gamma(k+1)} = \sum_{k=0}^{\infty} \frac{z^k}{k!} = e^z.$$

$$E_{1,m}(z) = \frac{1}{z^{m-1}} \left\{ e^z - \sum_{k=0}^{m-2} \frac{z^k}{k!} \right\} \quad (3.72)$$

The one-parameter function of the Mittag-Leffler type is

$$E_{\alpha}(z) = \sum_{k=0}^{\infty} \frac{z^k}{\Gamma(\alpha k + 1)} \quad (3.73)$$

The Laplace transform of the Mittag-Leffler function in two parameters is

$$\int_0^{\infty} e^{-t} t^{\beta-1} E_{\alpha,\beta}(zt^{\alpha}) dt = \frac{1}{1-z}, \quad (|z| < 1) \quad (3.74)$$

and the pair of Laplace transforms of the function $t^{\alpha k+\beta-1}E_{\alpha,\beta}^{(k)}(\pm z t^{\alpha})$ is:

$$\int_0^{\infty} e^{-pt} t^{\alpha k+\beta-1} E_{\alpha,\beta}^{(k)}(\pm a t^{\alpha}) dt = \frac{k! p^{\alpha-\beta}}{(p^{\alpha} \mp a)^{k+1}}, \quad (\mathrm{Re}(p) > |a|^{1/\alpha}) \tag{3.75}$$

Finally, Eq. (3.76) is

$$G_3(t) = \frac{1}{a} \sum_{k=0}^{\infty} \frac{(-1)^k}{k!} \left(\frac{c}{a}\right)^k t^{2p(k+1)-1} E_{2p-p,2p+pk}^{(k)} \left(-\frac{b}{a} t^{2p-p}\right)$$

$$= \frac{1}{a} \sum_{k=0}^{\infty} \frac{(-1)^k}{k!} \left(\frac{c}{a}\right)^k t^{2p(k+1)-1} E_{p,2p+pk}^{(k)} \left(-\frac{b}{a} t^p\right) \tag{3.76}$$

where $E_{\lambda,\mu}(z)$ is the Mittag-Leffler function in two parameters and its k-th derivative is given by

$$E_{\lambda,\mu}^{(k)}(t) = \frac{d^k}{dt^k} E_{\lambda,\mu}(t) = \sum_{j=0}^{\infty} \frac{(j+k)! t^j}{j! \Gamma(\lambda j + \lambda k + \mu)}, \quad k = 0, 1, 2, \ldots \tag{3.77}$$

The inverse Laplace transform of $G(s)$ is the fractional Green's function:

$$G(t) = {}_0D_t^p[G_3(t)] = \frac{d^p G_3(t)}{dt^p} \tag{3.78}$$

where the fractional derivative of $G_3(t)$, Eq. (3.76), is evaluated with the help of Eq. (3.66). The solution of the initial-value problem for the ordinary fractional linear differential equation with constant coefficients using Green's function is presented. The consideration taken into account for the given differential equation under homogeneous initial conditions is $b_k = 0$, $k = 1, 2, \ldots, n$:

$${}_aD_t^{\sigma_n} y(t) + \sum_{k=1}^{n-1} p_k(t) {}_aD_t^{\sigma_n-k} y(t) + p_n(t) y(t) = f(t),$$

$$\sigma_k = \sum_{j=1}^{k} \alpha_j, \ 0 \le \alpha_j \le 1, \ j = 1, 2, \ldots, n \tag{3.79}$$

The analytical solution of the given problem takes the form

$$y(t) = \int_0^t G(t, \tau) f(\tau) d\tau \tag{3.80}$$

where $G(t,\tau)$ is known as Green's function of Eq. (3.78); for the fractional differential equations with constant coefficients, this function is

$$G(t,\tau) = G(t-\tau) \tag{3.81}$$

and in such cases, Green's function can be obtained by the Laplace transform method. Therefore, the solution of linear fractional differential equations with constant coefficients reduces to finding the fractional Green's function. Finally, for nonhomogeneous initial conditions, the solution has the following form:

$$y(t) = \sum_{k=1}^{n} b_k \psi_k(t) + \int_0^t G(t-\tau) f(\tau) d\tau, \quad b_k = \left[{}_0 D_t^{\sigma_k - 1} y(t) \right]_{t=0} \tag{3.82}$$

At last, this allows an explicit representation of the solution:

$$U(t) = \int_0^t G(t-\tau) i(\tau) d\tau \tag{3.83}$$

Now again, the initial electromagnetic oscillation is represented by the differential equation, Eq. (3.5), and when the nonlinear terms are omitted and/or canceled the first step (i.e., homogeneous solution) may be obtained using a numerical calculation derived from the Grunwald definition [136], as is shown in Figure 3.7.

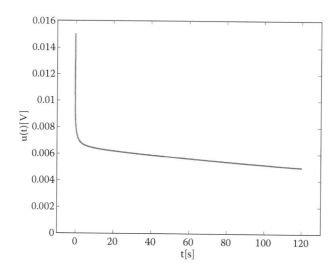

FIGURE 3.7

Calculated solution, Eqs. (3.61) and (3.62), homogeneous solution, after complete evaluation of linearized Eq. (3.5).

Note: The parameters are $\alpha = 0.9995$, $U_0 = 15$ mV, p = 0.95, and T = 0.001 s.

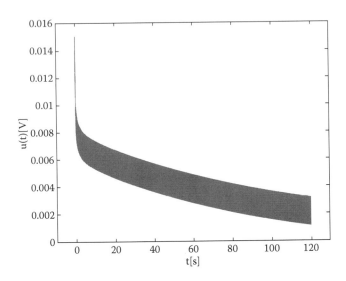

FIGURE 3.8

Calculated solution, Eqs. (3.61) and (3.62), nonhomogeneous solution, after complete evaluation of linearized Eq. (3.6).

Note: The parameters are $\alpha = 0.95$, $U_0 = 15$ mV, $p = 0.95$, $T = 0.01$ s, $A_n = 0.05$ nm, and $\cos(\omega_0 t) =$ $d^r/dt^r(\sin \omega_0 t)$.

The calculation has been done for the following parameters: $\alpha = 0.9995$, $U_0 = 15$ mV, $p = 0.95$, and $T = 0.001$ s.

Further on, considering Eq. (3.6), the nonnomogeneous solution is obtained, and presented in Figure 3.8. The obtained result appears as a band because the input (cos) is of the fractional order; and output is in a damped oscillatory mode of high frequencies! The calculation has been done for the following parameters [where the derivative of cos ($\omega_0 t$) is fractional too, of the order r]: $\alpha = 0.95$, $U_0 = 15$ mV, $p = 0.95$, $T = 0.01$ s, $A_n = 0.05$ nm, and $\cos(\omega_0 t) = d^r/dt^r (\sin \omega_0 t)$.

Here the presented model is derived only for the linearized van der Pol equation, while the more realistic nonlinear case will be the subject of the following section. It is recommended to perform numerical calculations using parallel data processing [1,136].

3.1.4.2.4 The van der Pol Fractional-Order Derivative Model: Nonlinear

Nonlinear fractional differential equations have received rather less attention in the literature, partly because many of the model equations proposed have been linear. Here will be considered a nonlinear homogeneous ($i(t) = 0$) differential equation of the van der Pol type which represents the droplet or droplet-film structure formation:

$$C\frac{dU}{dt} + \left(\frac{1}{R} - \alpha\right)U + \frac{1}{L}\int U\,dt + \gamma U^3 = 0 \qquad (3.84)$$

Also, one can obtain equivalent nonlinear problems applying a differentia-tion of Eq. (3.84) such as

$$C\frac{d^2U}{dt^2} + \left(\frac{1}{R} - \alpha + 3\gamma U^2\right)\frac{dU}{dt} + \frac{1}{L}U = 0 \tag{3.85}$$

In an effort to generalize the previous equation, the ordinary time deriva-tive and integral are now replaced with the corresponding fractional-order time derivative and integral of order p, or taking into account the Caputo definition, one can obtain the fractional-order van der Pol equation:

$$_0^C D_t^{2p}U(t) = -\frac{1}{C}\left(\frac{1}{R} - \alpha\right)_0^C D_t^p U(t) - \frac{3\gamma}{C}U(t)^2 {}_0^C D_t^p U(t) - \frac{1}{CL}U(t) \tag{3.86}$$

Now, one may convert previous equations with commensurate multiple fractional derivatives into an equivalent system of equations of low order. Let

$$x_1(t) = U(t), \quad x_2(t) = {}_0^C D_t^p U(t), \quad p \in Q \tag{3.87}$$

In that way, introducing vector $x(t) = (x_1, x_2)^T$, one can get in condensed form

$$_0^C D_t^p x(t) = \begin{bmatrix} 0 & 1 \\ -1/LC & -(1/R - \alpha)/C \end{bmatrix} \begin{Bmatrix} x_1(t) \\ x_2(t) \end{Bmatrix}$$
$$+ \begin{bmatrix} 0 & 0 \\ 0 & -3\gamma x_1^2(t)/C \end{bmatrix} \begin{Bmatrix} x_1(t) \\ x_2(t) \end{Bmatrix} \tag{3.88}$$

or

$$_0 D_t^p x(t) = Ax(t) + B(x_1(t))x(t) \tag{3.89}$$

It is easily observed that the previous case is one general case for this non-linear problem, which can be obtained in the following form:

$$_0^C D_t^p x(t) = f(t, x(t)) \tag{3.90}$$

subject to the initial conditions:

$$x_k(0) = \begin{cases} x_0^{(kp)}, & kp \in N_0 \\ 0, & else \end{cases} \tag{3.91}$$

Hence, the problem of finding a unique continuous solution of Eqs. (3.90) and (3.91) is omitted (for more details, see [133,134]). Now a case of low fractionality is considered [127], that is, the order of fractional derivative p slightly deviates from an integer value n ($p = n - \delta$, $n = 1,2$, $\delta \ll 1$):

$$_0^\epsilon D_t^{2-\delta} U(t) = -\frac{1}{C}\left(\frac{1}{R} - \alpha\right) {_0^\epsilon} D_t^{1-\delta} U(t) - \frac{3\gamma}{C} U(t)^2 {_0^\epsilon} D_t^{1-\delta} U(t) - \frac{1}{CL} U(t), \quad (3.92)$$

Introducing the Caputo fractional derivative in the form (3.93), when in the limit $\varepsilon \to 0$

$$_0 D_t^p f(t) = \frac{f^{(n)}(0) t^{n-p}}{\Gamma(n-p+1)} + \frac{1}{\Gamma(n-p+1)} \int_0^t f^{(n+1)}(\tau)(t-\tau)^{n-p} d\tau, \quad n-1 < p \le n, \quad (3.93)$$

one can obtain

$$_0 D_t^{n-\delta} f(t) = \frac{f^{(n)}(0) t^\delta}{\Gamma(1+\delta)} + \frac{1}{\Gamma(1+\delta)} \int_0^t f^{(n+1)}(\tau)(t-\tau)^\delta d\tau, \quad n=1,2 \quad (3.94)$$

First, considering the case $\delta t \ll 1$ and using the expansion

$$\frac{1}{\Gamma(1+\varepsilon)}(t-\tau)^\varepsilon = \frac{1}{\Gamma(1+\varepsilon)} e^{\varepsilon \ln(t-\tau)} \quad (3.95)$$

this yields the fractional derivative in a form of the perturbation to the second derivative and first derivative:

$$_0 D_t^{2-\delta} f(t) = f^{(2)}(t) + \delta\left(f^{(2)}(0)\ln t + \gamma f^{(2)}(t) + \int_0^t f^{(3)}(\tau)\ln(t-\tau)d\tau \right) + \ldots, \quad (3.96)$$

$$_0 D_t^{1-\delta} f(t) = f^{(1)}(t) + \delta\left(f^{(1)}(0)\ln t + \gamma f^{(1)}(t) + \int_0^t f^{(1)}(\tau)\ln(t-\tau)d\tau \right) + \ldots, \quad (3.97)$$

where $\tau < t$, $\gamma = 0.5772156649\ldots$ is a constant. As a result, when the limit $\delta \to 0$, the correct expansion is

$$\lim_{\delta \to 0} {_0} D_t^{n-\delta} f(t) = f^{(n)}(t), \quad n=1,2 \quad (3.98)$$

and one can obtain Eq. (3.58). Now, the asymptotic representation of the fractional derivative, using the Laplace transform and its inversion, expansion $\delta t \gg 1$, can be formally obtained as follows:

$$_0D_t^p f(t) = \frac{-f(0)t^{-p}}{\Gamma(1-p)} + \frac{-f'(0)t^{1-p}}{\Gamma(2-p)} + \sum_{k=0}^{\infty}\frac{F^{(k)}(0)t^{-p-k-1}}{\Gamma(-p-k)k!}, \quad 1 < p < 2, (t \to \infty) \quad (3.99)$$

$$_0D_t^p f(t) = \frac{-f(0)t^{-p}}{\Gamma(1-p)} + \sum_{k=0}^{\infty}\frac{F^{(k)}(0)t^{-p-k-1}}{\Gamma(-p-k)k!}, \quad 0 < p < 1, (t \to \infty) \quad (3.100)$$

where $F(s)$ is a Laplace transform of $f(t)$. Taking $p = n - \delta$, we obtain

$$_0D_t^{2-\delta} f(t) \approx \frac{-f(0)t^{-2+\delta}}{\Gamma(-1+\delta)} - \frac{f'(0)t^{-1+\delta}}{\Gamma(\delta)} \approx \delta f(0)t^{-2+\delta} - \delta f'(0)t^{-1+\delta}, \quad (\delta t \gg 1) \quad (3.101)$$

$$_0D_t^{1-\delta} f(t) \approx \frac{-f(0)t^{-1+\delta}}{\Gamma(-1+\delta)} \approx \delta f(0)t^{-1+\delta}, \quad (\delta t \gg 1) \quad (3.102)$$

After changing the fractional derivatives with Eqs. (3.101) and (3.102), Eq. (3.92) becomes

$$-\frac{3\gamma}{C}U^2(t)(\delta U(0)t^{-1+\delta}) - \frac{1}{CL}U(t) = \delta U(0)t^{-2+\delta} - \delta U'(0)t^{-1+\delta}$$

$$+ \frac{1}{C}\left(\frac{1}{R} - \alpha\right)(\delta U(0)t^{-1+\delta}) \quad (3.103)$$

In condensed form:

$$a_1(t)U^2(t) + a_2(t)U(t) = a_0(t) \quad (3.104)$$

where

$$a_1(t) = -\frac{3\gamma}{C}(\delta U(0)t^{-1+\delta}), \quad a_2 = -\frac{1}{LC},$$

and

$$a_o(t) = \delta t^{-1+\delta}(U(0)t^{-1} - U'(0)) + \frac{1}{C}\left(\frac{1}{R} - \alpha\right)(\delta U(0)t^{-1+\delta})$$

Solving this quadratic equation, one can see the asymptotic behavior $U(t) = U(\delta, t^{-1+\delta}, t^{-2+\delta})$ when $t \to \infty$.

3.1.4.2.5 The Numerical Method-Predictor-Corrector Algorithm

Recently, there has been a growing interest to develop an approximate numerical technique for fractional differential equations (FDEs) which are numerically stable and which can be applied to both linear and nonlinear FDEs. In some cases for a numerical calculation of nonlinear FDEs, one can use the fact that a fractional derivative is based on a convolution integral, the number of weights used in the numerical approximation to evaluate fractional derivatives. Hence, it is suggested to apply a predictor-corrector algorithm for the solution of systems of nonlinear equations of lower order. The algorithm is a generalization of the classical Adams-Bashfourh-Moulton algorithm that is well known for the numerical solution of first-order problems [122]. This approach is based on rewriting the initial value problem, Eqs. (3.90) and (3.91), as an equivalent fractional integral equation (a Volterra integral equation of the second kind):

$$x'(t) = x'_0 + \frac{1}{\Gamma(p)} \int_0^t f'(s, x'(s))(t-s)^{p-1} \, ds, \tag{3.105}$$

The proposed algorithm presents a scalar scheme, and no problems at all, since simple application of the previous algorithm in a component-wise fashion is performed to each component of the vector problem. When one needs to take into account the effects like the round-off and truncation error, it is possible to ask for a step size h, whose combined effect arises from both error sources and is minimized. The truncation error decreases with a step size h, while round-off tends to have the opposite behavior. Roughly speaking, an optimum step size would be of the order of $h \sim \sqrt{\varepsilon_x x(s)/\ddot{x}(p)}$, where $s, p \in [0, t]$, s, and p are some numbers within the interval $(0, t)$ [40], and ε_x is a relative accuracy with which one can compute x. Also, numerical schemes that are *convergent*, *consistent*, and *stable* are of interest, but these properties are not treated here. A numerical algorithm that solves Caputo-type fractional differential equations is listed below, $(0 < p < 2, p \neq 1)$. So, one may introduce the uniformly distributed grid points for the time interval $[0, t = X]$ $t_k = kh$, $k = 0, 1, 2, \ldots, N : h = X / N$.

Step 1. Predict using the quadrature weights (derived from a product rectangular rule) [36].

$$x_h^p(N) = \sum_{m=0}^{[p]} \frac{x^m}{l!} x_{0+}^m + \left[\frac{h^p}{\Gamma(1+p)} \right] \sum_{k=0}^{N-1} b_{K,N} f(t_k, x_k) \tag{3.106}$$

where

$$b_{k,N} = (N-k)^p - (N-k-1)^p \tag{3.107}$$

Step 2. Evaluate $f(X, x_N^P = x_h^P(N))$.

Step 3. Correct with the following expression where the quadrature weights (derived from the product trapezoidal rule)

$$x_h(N) = \sum_{m=0}^{[p]} \frac{X^m}{l!} x_{0^+}^{(m)} + \left[\frac{h^p}{\Gamma(2+p)} \right] \left(\sum_{k=0}^{N-1} c_{k,N} f(t_k, x_k) + c_{N,N} f(X, x_N^P) \right) \tag{3.108}$$

where

$$c_{k,N} = \begin{cases} (1+p)N^p - N^{1+p} + (N-1)^{1+p}, & k=0, \\ (N-k+1)^{1+p} - 2(N-k)^{1+p} + (N-k-1)^{1+p}, & 0 < k < N, \\ 1, & k=N, \end{cases} \tag{3.109}$$

Step 4. Reevaluate $f(X, x_N = x_h(N))$, and save it as $f(t_N, x_N)$, which is then used in the next integration step. Finally, it is said to be of the *predict, evaluate, correct, evaluate* (PECE) type because, in this implementation, start run by calculating the predictor in Eq. (3.106); then evaluate $f(X, x_N^P = x_h^P(N))$, and use this to calculate the corrector in Eq.(3.108); and, finally, evaluate $f(t_N, x_N)$.

3.1.4.2.5.1 Solution of the Representative Nonlinear Model: A Homogeneous Case The initial electromagnetic oscillation is represented by the nonlinear fractional differential Eq. (3.5); a homogeneous solution may be obtained using a numerical calculation of the Caputo derivative and PECE algorithm, Eqs. (3.106)–(3.109), as is shown in Figure 3.9. The calculation has been done for the following parameters: $\alpha = 8 \times 10^{-7}$, $Uo = 8mV$, $p = 1.2$, $T = 0.004\,s$, and $\gamma = 3 \times 10^{-3}$.

3.1.4.2.5.2 Solution of the Representative Nonlinear Model: A Nonhomogeneous Case The particular incident electromagnetic oscillation is represented by the nonlinear fractional differential Eq. (3.6); a nonhomogeneous solution may be obtained using a numerical calculation of the Caputo derivative and PECE algorithm, Eqs. (3.106)–(3.109), as is shown in Figure 3.10. The calculation has been done for the following parameters:

$$[\alpha = 8 \cdot 10^{-7}, Uo = 8mV, p = 1.2, T = 0.004\,s, \gamma = 3 \cdot 10^{-3}$$

$$\omega_1 = 1.2 \times 10^8, A = 1 \times 10^{-10}]$$

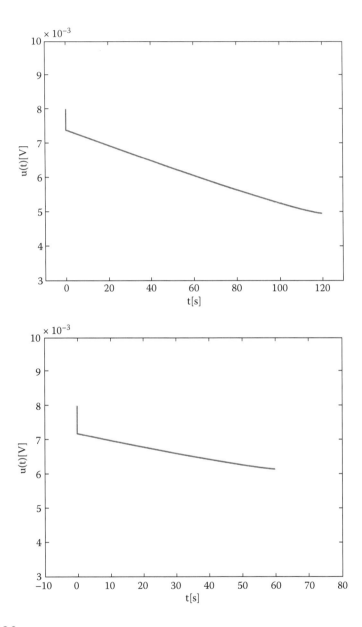

FIGURE 3.9

Calculated solution, (a) Eqs. (3.106)–(3.109), homogeneous solution, after complete evaluation of nonlinear Eq. (3.5).

Note: The parameters are $\alpha = 8 \times 10^{-7}$, $U_0 = 8$ mV, $p = 1.2$, $T = 0.004$ s, and $\gamma = 3 \times 10^{-3}$; and (b) detail.

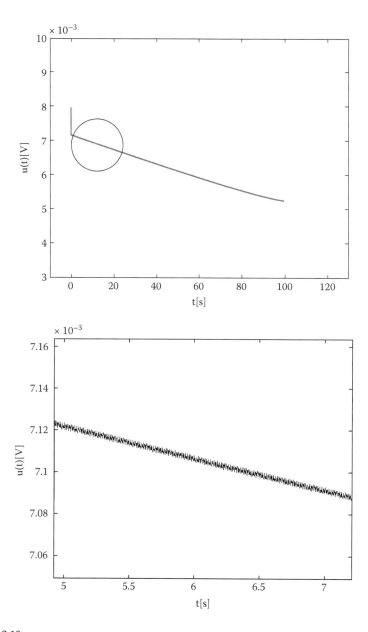

FIGURE 3.10

Calculated solution, Eqs. (3.106)–(3.109), nonhomogeneous solution, after complete evaluation of nonlinear Eq. (3.6).

Note: The parameters are $\alpha = 8 \times 10^{-7}$, $U_0 = 8$ mV, $p = 1.2$, $T = 0.004$ s, $\gamma = 3 \times 10^{-3}$, $\omega = 1.2 \times 10^{8}$, and $A = 1 \times 10^{-10}$; and (b) Detail.

Appendix

Here, the problem of finding a unique continuous solution of Eqs. (3.92) and (3.93) is considered. Using some classical results from fractional calculus, that is, applying the following lemma, one can obtain a solution of the initial-value problem of Eqs. (3.92) and (3.93) [124].

> **Lemma 1** Let the function f be continuous. The function $x(t)$ is a solution of the Cauchy problem, Eqs.(3.103) and,(3.104), if and only if
>
> $$x(t) = \sum_{j=0}^{n-1} \frac{t^j}{j!} x^{(j)}(0) + \frac{1}{\Gamma(p)} \int_0^t f(s, x(s))(t-s)^{p-1} ds, \quad n-1 < p \le n \qquad (3.94)$$

This equation is weakly singular if $0 < p < 1$ and regular for $p \ge 1$. In the former case, we must give explicit proofs for the existence and uniqueness of the solution. Hence, following results that are very similar to the corresponding classical theorems of existence and uniqueness, known in the scalar case of first-order equations, are the following [123]:

> **Theorem 1 (Existence)** Assume that $D:[0,T^{\cdot}] \times [x_0^{(0)} - c, x_0^{(0)} + c]$ with some $T^{\cdot} > 0$ and some $c > 0$, and let the function $f:D \to \mathbb{R}$ be continuous. Furthermore, define $T = \min\{T^{\cdot}, (c\Gamma(p+1)/\|f\|_\infty)^{1/p}\}$. Then, there exists a function $x:[0,T] \to \mathbb{R}$ solving the Cauchy problem, Eqs. (3.103) and (3.104), scalar case.

> **Theorem 2 (Uniqueness)** Assume that $D:[0,T^{\cdot}] \times \left[x_0^{(0)} - c, x_0^{(0)} + c \right]$ with some $T^{\cdot} > 0$ and some $c > 0$. Also, let the function $f:D \to \mathbb{R}$ be bounded on D and fulfill a Lipschitz condition with respect to the second variable, that is,
>
> $$|f(t,\tilde{x}) - f(t,\bar{x})| \le \lambda |\tilde{x} - \bar{x}| \qquad (3.95)$$
>
> with some constant $\lambda > 0$ independent of $t, \bar{x}, \tilde{x}$. Then, there exists at most one function $x:[0,T] \to \mathbb{R}$ solving the Cauchy problem, Eqs. (3.92) and (3.93), scalar case.

The generalization of previous theorems to vector-valued functions x is immediate. The proof of the uniqueness theorem will be based on the generalization of Banach's fixed-point theorem [124].

> **Remark 1:** Without the Lipschitz assumption on f, the solution need not be unique.

4

Experimental

4.1 Experimental

4.1.1 Experimental Confirmation

4.1.1.1 Description of the Physical-Chemical System

The particular secondary liquid-liquid system which has been used to cor-
roborate the validity of the theoretical predictions was a heavy-phase-droplet
and light-phase-film structure immersed in a heavy-phase continuum (double emulsion). This system was the heavy-phase output from a "pump-mix"
mixer-settler battery together with its entrained light phase. The battery was
a part of a pilot plant for the extraction of uranium from wet phosphoric
acid by the D2EHPA-TOPO process [28,33,36–45,47–52,58,62,92,97,101]. The
heavy liquid was 5.6 M phosphoric acid, and the light liquid was the synergistic mixture of 0.5 M di(2-ethylhexyl) phosphoric acid and 0.125 M tri-n-
octylphosphine oxide in dearomatized kerosene [38,48–52].

The structural formulae of the constituent liquids are

$$
\begin{array}{ccc}
\text{OH} & \text{OH} & \text{R}' \\
| & | & | \\
\text{HO} - \text{P} = \text{O} & \text{RO} - \text{P} = \text{O} & \text{R}' - \text{P} = \text{O} \\
| & | & | \\
\text{OH} & \text{OR} & \text{R}'
\end{array}
$$

where R is CH_2-$CH(CH_2)_3CH_3$, and R' is $(CH_2)_7CH_3$.
$$
\begin{array}{c}
| \\
C_2H_5
\end{array}
$$

4.1.1.2 Generation of the Physical Model

A polydispersion was generated and the primary separation performed in
a laboratory mixer for the batch studies under the conditions applied in the

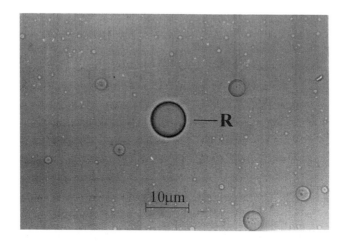

FIGURE 4.1

Photograph *in situ* of the examined liquid-liquid droplet-film structure submerged into the droplet homophase continuum.

Note: R: rigid spheres.

pilot plant [28,62]. The selected hydrodynamic characteristics in the mixer unit were as follows:

- The ratio of the phases in the mixer, light:heavy, was equal to 1.1.
- The number of revolutions of the eight-blade double-shrouded impeller in the mixer was equal to 15 s^{-1}.
- The mixing criterion, $\rho n^3 D^2$, was equal to 13.

Thereafter, the sample of the heavy phase, together with the entrained light phase, was isolated and observed in situ with an optical microscope. Figure 4.1 shows an *in situ* photograph of the examined liquid-liquid droplet-film structure immersed in the droplet homophase continuum. This isolated system is in hydrodynamic and electrodynamic equilibrium (i.e., the rigid sphere can be observed).

Further, Figure 4.2 shows an *in situ* photograph of the system examined at the junction point of the droplet-film structure and plate. The glass plate disturbs this system during observation (i.e., the elastic sphere can be observed).

Finally, Figure 4.3 shows an *in situ* photograph where both rigid and elastic spheres may be observed.

4.1.1.3 Measuring the Electrical Interfacial Potential

A method and apparatus were developed to monitor voltammetrically the electrical interfacial potential (EIP) appearing during the formation of the electric double layer (EDL) while the two-phase contact occurs [11,51,62]. Measurements of the EIP, including jump, have been performed during the

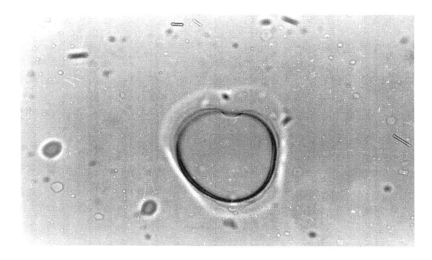

FIGURE 4.2
Photograph *in situ* of the examined system at the droplet-film structure-plate junction point.

Note: E: elastic sphere, disturbed.

Source: Reprinted from Spasic, A. M., Jokanovic, V., Krstic, D. N., A theory of electrovis-coelasticity: A new approach for quantifying the behavior of liquid-liquid interfaces under applied fields, *J. Colloid Interface Sci.*, 186, 1997, p. 437, with permission from Academic Press.

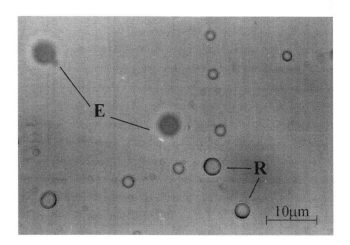

FIGURE 4.3
Photograph *in situ* of the examined system.

Note: R: rigid spheres; and E: elastic spheres, disturbed ("breathing" expansion and contraction).

Source: Reprinted from Spasic, A. M., Jokanovic, V., Krstic, D. N., A theory of electroviscoelastic-ity: A new approach for quantifying the behavior of liquid-liquid interfaces under applied fields, *J. Colloid Interface Sci.*, 186, 1997, p. 437, with permission from Academic Press.

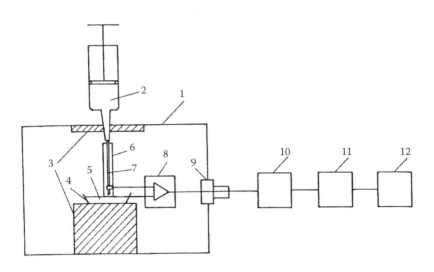

FIGURE 4.4
The liquid-liquid contact cell with its peripheral modulus.

Note: (1) The Faraday cage; (2) the syringe with the heavy phase; (3) the insulators; (4) the platinum vessel; (5) the light phase; (6) the plastic tube; (7) the steel needle; (8) the high-input-impedance instrumental amplifier; (9) the connector with coaxial cable; (10) the oscilloscope with memory; (11) the analogue to digital converter; and (12) the data acquisition system.

processes of formation and transition of the electroviscoelastic sphere into the rigid sphere [51]. Figure 4.4 shows the developed liquid-liquid contact cell with its peripheral modulus for the EIP measurements.

Measurements of the EIP, including jump, during the EDL formation at the boundary interface of two liquids were performed by introducing the heavy liquid through a syringe, whose needle constituted one electrode, into a platinum vessel, which constituted the other electrode. The electrodes were connected via a sensor (high-input-impedance instrumental amplifier) to an oscilloscope with memory. A Faraday cage, to avoid effects of the environment, surrounded the electrodes and the sensor. Figure 4.5 shows a diagram of the high-input-impedance instrumental amplifier that was built for this purpose.

During the liquid-liquid contact cell development, two important arguments were considered: the first concerned very low energy levels appearing during the EDL formation at the boundary surface of two immiscible liquids, and the second concerned the possible influence of the cell construction elements on the impedance structure determination.

4.1.1.4 Measuring the Resonant Frequency

In order to determine the resonant (characteristic) frequency, the nuclear magnetic resonance (NMR) spectrometer was used as a reactor for the

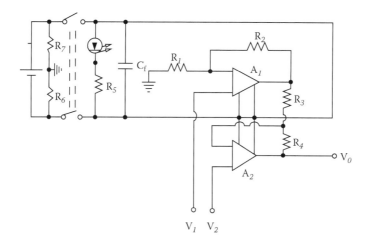

FIGURE 4.5

A schematic diagram of the constructed high-input-impedance instrumental amplifier.

Note: $R_1 - R_4$: the matched resistances; C_f: the capacitance; A_1 and A_2: the high-input-impedance instrumental amplifiers; and V_0, V_1, and V_2: the input and output potentials.

energetic analysis. The impedance Z at the resonant frequency ω_0 is equal to the resistance R. The resonant frequency of the electromechanical oscillator can be considered as some characteristic frequency within the vibro-rotational spectrum of the molecular complex that builds the droplet-film structure [11,51].

All experiments were performed and all spectra acquired on a Bruker MSL 400 spectrometer with a 9.395 T magnet and at a ^{31}P frequency of 161.924 MHz. The transmitter was set at resonance frequency with the phosphoric acid standard solution, and a sweep width of 15000 Hz was employed. The swept region corresponded to the range between –10 and 90 ppm.

4.1.2 Results and Discussion

4.1.2.1 General Reminder

The droplets or droplet-film structures with diameters in the range between the two critical diameters, d_1 and d_2 (micro) and d_2 and d_3 (nano), consist of a finite number of electromechanical oscillators/clusters. Droplets with diameters smaller than the third critical diameter, d_3, are formed by a very large (almost infinite) number of entities or atto clusters.

The stability of the interface or its electroviscoelastic behavior is specified in terms of three forms: these are rigid, elastic, and plastic, as is shown in Figures 3.2 and 4.6 and 4.10.

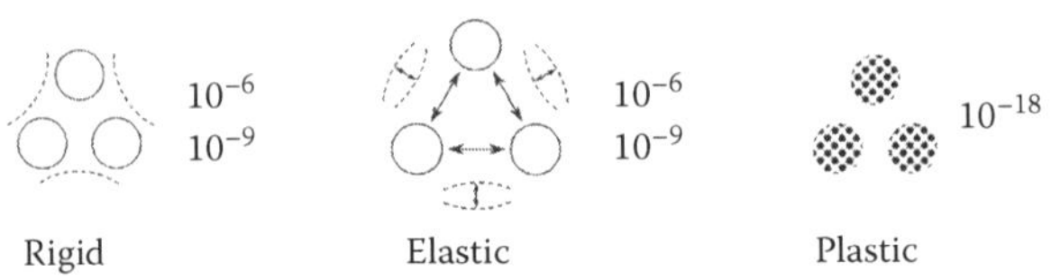

Rigid   Elastic   Plastic

FIGURE 4.6
The electroviscoelastic behavior: rigid, elastic, and plastic.

Source: Reprinted from Spasic, A. M., Jokanovic, V., Krstic, D. N., A theory of electroviscoelasticity: A new approach for quantifying the behavior of liquid-liquid interfaces under applied
fields, *J. Colloid Interface Sci.*, 186, 1997, p. 443, with permission from Academic Press.

The essential equation of a general theory of electroviscoelasticity can be
expressed by

$$F_d(V) = F_d(V_j) + [F_d(V_s) + F_d(V_b)] \tag{4.1}$$

The first term on the right-hand side of the equation is due, for example, to
the energy effects, and the second term (consisting of two subterms) is due
to the entropic effects. From the second, entropic term, the separation factors,
internal S_i and external S_e, can be derived and can be used as main quantities
characterizing the dispersed system.

4.1.2.2 Emulsions as the Particular Case

4.1.2.2.1 EIP Measurements

Figure 4.7 shows the measured change in EIP appearing during the introduction of the heavy-phase droplet into the light-phase continuum [6]. It can
be seen in the figure that an interfacial jump potential peak appears during
the formation of the EDL. Thereafter, the change of the EIP decreases to a
constant value! The lowering of the EIP in absolute value during the flow is
due to the participation of cations that form the dense part of the EDL. The
anions are the counterions in the diffuse part. Redistribution of these anions
and cations between near-surface and surface layers of the heavy phase
defines the kinetics of the EIP [11,51].

Figure 4.8 shows measured spontaneous oscillations of the EIP during the
"breathing" period. After the EIP jump, which is in the millivolt-millisecond
scale, the EIP continues to oscillate in the millivolt-minute range. Its damped
oscillatory mode is (probably) due to the hydrodynamic instability of the
interfacial surface, as a consequence of the local gradients of interfacial tension
and density in mutual saturation processes of liquids [11,51]. Other relevant
interpretations of the EIP spontaneous oscillations may be expressed as follows: the electroviscoelastic sphere undergoes transformation into the rigid
sphere. This transformation process can be understood as memory storage.

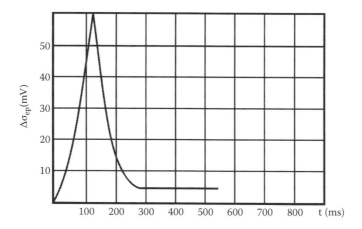

FIGURE 4.7

Measured variations of the EIP with time for the examined system of phosphoric acid and D2EHPA-TOPO-kerosene at the spherical interface.

Source: Reprinted from Spasic, A. M., Jokanovic, V., Krstic, D. N., A theory of electroviscoelasticity: A new approach for quantifying the behavior of liquid-liquid interfaces under applied fields, *J. Colloid Interface Sci.*, 186, 1997, p. 441, with permission from Academic Press.

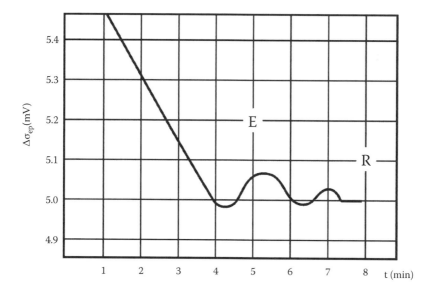

FIGURE 4.8

Measured spontaneous oscillations of the EIP during the "breathing" period: Transformation of the electroviscoelastic sphere into the rigid sphere.

Source: Reprinted from Spasic, A. M., Jokanovic, V., Krstic, D. N., A theory of electroviscoelasticity: A new approach for quantifying the behavior of liquid-liquid interfaces under applied fields, *J. Colloid Interface Sci.*, 186, 1997, p. 441, with permission from Academic Press.

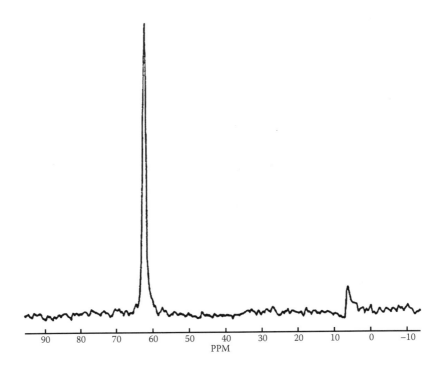

FIGURE 4.9

^{31}P NMR spectrum of a D2EHPA-TOPO-kerosene sample (phosphoric acid standard solution, sweep width of 15000 Hz).

Note: A peak at 7 ppm corresponds to D2EHPA, and a peak at 62 ppm corresponds to TOPO.

Source: Reprinted from Spasic, A. M., Jokanovic, V., Krstic, D. N., A theory of electroviscoelasticity: A new approach for quantifying the behavior of liquid-liquid interfaces under applied fields, *J. Colloid Interface Sci.*, 186, 1997, p. 437, with permission from Academic Press.

4.1.2.2.2 ^{31}P NMR Measurements

Figure 4.9 shows ^{31}P NMR spectra of the molecular complex that builds the examined droplet-film structure. The impedance Z at the resonant frequency ω_0 is equal to the resistance R. The resonant frequency of the electromechanical oscillator can be considered as some characteristic frequency within the vibrorotational spectrum of the molecular complex that forms the droplet-film structure.

4.1.3 Assembled Measured, Calculated, and Estimated Data

Finally, the measured, calculated, and estimated data relevant to the developed theory are given in Table 4.1 [38,48].

The critical diameters d_1 and d_2 of the droplets and droplet-film structures were determined by optical microscope (Figures 4.1, 4.2, and 4.3). Some of the

TABLE 4.1

Measured, Calculated, and Estimated Data Relevant to the Developed Theory

Address	Data	Unit	Value	Source[*]
1	Critical diameter, d_1	μm	8	M; R51; R62; Fig 4.1
2	Critical diameter, d_2	μm	1	M; R51; R62; Fig 4.1
3	Representative diameter, d_{32}	μm	6.35	C; R51; R62
4	Density H$_3$PO$_4$, ρ	kgm^{-3}	1300	M; R51; R48; at 313 K; Fig 2.2
5	Density DTK, ρ	kgm^{-3}	800	M; R51; R48; at 313 K; Fig 2.2
6	Viscosity H$_3$PO$_4$, μ	Pa s	0.029	M; R51; R62; at 313 K; Fig 2.2
7	Viscosity DTK, μ	Pa s	0.019	M; R6; R10; at 313 K; Fig 2.2
8	Interfacial tension, σ_{in}	mNm^{-1}	16.4	M; R51; R62; at 313 K; Fig 2.2
9	EIP pik, σ_{ep}	mV	60	M; R51; R62; Fig. 4.7
10	EIP relaxed, σ_{ep}	mV	5	M; R51; R62; Figs. 4.7, 4.8
11	Consequent current, I	nA	60	C; R51; R48
12	Impedance, Z_d	MΩ	1	C; R51; R7; (Z = σ_{ep}/I)
13	Dielectric constant, ε	Fm^{-1}	24×10^{-12}	T; R20
14	Capacitance, C_d	pF	24	C; R51; R48
15	Inductance, L_d	μH	4.7	C; R51; R48; L = $(2\pi f_0)^{-2}C^{-1}$
16	Resonant frequency, ω_0	rads^{-1}	120×10^7	C; R51; Eq.3.12; Fig. 3.5
17	Resonant frequency, f_0	Hz	$184{,}1 \times 10^6$	C; R51
18	Resonant frequency H$_3$PO$_4$, f_0	Hz	161.924×10^6	M; R51; Fig. 4.9
19	Resonant frequency D2EHPA, f_0	Hz	161.925×10^6	M; R51; Fig. 4.9
20	Resonant frequency TOPO, f_0	Hz	161.934×10^6	M; R51; Fig. 4.9
21	Mechanical wave energy, w_m	Jm^{-3}	137.6	C; R51; Eq. 3.27
22	Electrical wave energy, w_e	Jm^{-3}	256	C; R51; Eqs. 3.15–3.17
23	Free surface energy, ΔG	Jm^{-3}	200	C; R51; Eq. 3.3
24	Initial amplitude, ξ_0	nm	1.5	C; E; R51; Eq. 3.23
25	Initial amplitude, ξ_0	nm	1.7	C; M; R51; Eq. 3.23
26	Internal separation, S_i	nm	1.5	C; E; Fig. 3.1
27	External separation, S_e	nm	0.3	C; E; R22; Fig. 3.1
28	Depth of DTK fim, δ_{32}	nm	4.91	C; R51; R48

[*] C = calculated; E = estimated; M = measured; R = reference; T = table.

Source: Spasic, A. M., Jokanovic, V., Krstic, D. N., A theory of electroviscoelasticity: A new approach for quantifying the behavior of liquid-liquid interfaces under applied fields, *J. Colloid Interface Sci.*, 186, 434–446, 1997. Reprinted by permission of Academic Press.

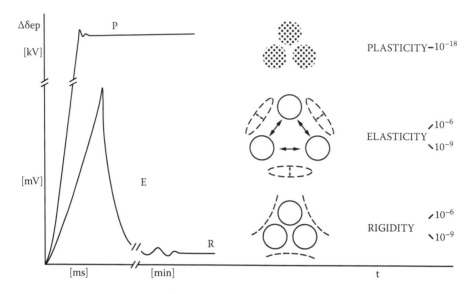

FIGURE 4.10
Correlation of the interfacial electrical potential and time: Behavior of micro-, nano-, and atto-dispersed systems.

physical and physicochemical properties were determined using the classical methods, ρ, μ, and σ_{in}. Further on, the electrical interfacial potential σ_{ep}, including the jump, was monitored using the developed liquid-liquid contact cell (Figures 4.4 and 4.5). Finally, the resonant and characteristic frequency was determined using the NMR spectrometer, explained in Section 4.1.1.4 of this chapter. Other data, which are given in Table 4.1, were calculated or estimated using the adequate equations—eqs. (3.3), (3.15), (3.34), and (3.66)—some of which are presented in the table. In the last column, "Source," it is shown how and where the data were obtained.

It can be seen from this table, when comparing the experimental and calculated or estimated data, that the theoretical predictions are in fair agreement with the experimental results.

A graphical interpretation of the theoretical and experimental contents discussed in this book can be schematically presented in Figure 4.10. This figure shows the correlation, the change of the interfacial electrical potential with time, as well as the consequent behavior of micro- and nanodispersed systems, which may behave as rigid and/or elastic, and atto-dispersed systems, which behave only as plastic.

5

Conclusions

5.1 Implications

5.1.1 The First Philosophical Breakpoint

5.1.1.1 Application to the Particular Entrainment Problem in Solvent Extraction: Breaking of Emulsions

This chapter is an attempt to present some of the practical and engineering utility potentials of the topics that were discussed in Chapter 2, Sections 2.1 and 2.1.1 and 2.1.2. Therefore, some of the electromechanical principles presented here have been used for the elucidation of the secondary liquid-liquid phase separation problems, methods, equipment, and/or plant conception. The example given is related to the appearance of droplets and emulsions, or droplet-film structures and double emulsions, in solvent extraction operations. As these emulsions or double emulsions occur as an undesirable side effect, both nondestructive and destructive methods for their separation or elimination can be shown. Table 5.1 presents the two kinds of methods for the secondary liquid-liquid phase separation [11,62].

The particular problem considered as the representative one was mechanical entrainment of one liquid phase by the other in the solvent extraction operation. Experimental results obtained in the pilot plant for uranium recovery from wet phosphoric acid were used as the comparable source [27–45,51,62].

In this pilot plant, a secondary liquid-liquid phase separation loop was also carried out [38–40]. The loop consisted of a lamellar coalescer and four flotation cells in series. Central equipment in the loop, relevant to this investigation, is the lamellar coalescer. The phase separation in this equipment is based on the action of external forces of mechanical and/or electrical origin, while adhesive processes at the inclined filling plates occur [38,41,45,51].

A complete evaluation of the knowledge and database organization system for this particular problem in solvent extraction can be found in reference [62].

TABLE 5.1

The Secondary Liquid-Liquid Phase Separation Methods

Methods → Considerations ↓	Coalescence Lamellar	Coalescence Densely Packed	Flotation Aero	Flotation Air Induced	Electrostatic	Electrodynamic	Centrifugal	Dissolution	Demulsifier
Process									
Efficiency (%)	60–70	90	—	70–80	80–95	—	—	—	—
Sensitivity (%)	0.01	0.002	—	0.005	0.001	—	—	—	—
Operational									
Throughput	Big	Small	Small	Big	Small	—	Small	—	—
Flexibility	Yes	No	No	Yes	No	—	No	Yes	Yes
Maintenance	Easy	Difficult	Difficult	Easy	Difficult	—	Difficult	Easy	Easy
Specific	— — —	STC TSP SFC	STC TSP SFC	STC TSP SFC	— TSP —	—	— — —	—	—
Design									
Information	Insufficient	Insufficient	Insufficient	Lot	Insufficient	No	Insufficient	Insufficient	Insufficient
Patents									
Economic									
Capital Costs	Low	High	Medium	Very High	Very High	Very High	Low	Low	Low
Rentability	Medium	High	—	Medium	Low	Low	High	High	High

Note: STC: sensitive to the temperature changes; TSP: sensitive to the presence of the third stable phase; SFC: sensitive to the feed content of the entrained phase.

Source: Spasic, A. M., Djokovic, N. N., Babic, M. D., Marinko, M. M., Jovanovic, G. N., Performance of demulsions: Entrainment problems in solvent extraction, *Chem. Eng. Sci.*, 52, 657–675, 1997. Reprinted with permission from Elsevier Science.

5.1.1.2 Other Applications

A theory of electroviscoelasticity using fractional approach constitutes a new interdisciplinary approach to colloid and interface science. Hence, more degrees of freedom are in the model, memory storage considerations and hereditary properties are included in the model, and history impact to the present and future is in the game!

This theory and models discussed may facilitate the understanding of, for example, very developed interfaces in colloid and interface science, chemical and biological sensors, electro-analytical methods, and biology/biomedicine (hematology, genetics, and electroneurophysiology).

Furthermore, both the model and theory may be implemented in studies of structure; interface barriers/symmetries – surface (bilipid membrane cells, free bubbles of surfactants, Langmuir Blodget films), – line (genes, liquid crystals, microtubules), – point (fullerenes, micro-emulsions) and – overall (dry foams, polymer elastic, and rigid foams) [1, 11, 137–147].

5.1.2 The Second Philosophical Breakpoint

Also, the further evaluation of the idea related to the entity understood as an energetic ellipsoid based on the model of electrons following Maxwell-Dirac Isomorphism MDI seems to be sensible. This idea is directed toward one important question of the classical limit of quantum mechanics, i.e., is quantum mechanics applicable at a macroscopic level? This question was the result of research of complex systems by the end of the last century. According to the developed strategy the proposition appeared as: if the macroscopic physical systems are only the special case of a quantum-mechanical systems (e.g., like in von Neumann's theory of the measurement problem) then it is possible to observe under specified conditions, their quantum mechanical behavior.

The behavior of a droplet-film structure submerged into the droplet homophase or double emulsion, including its formation-existence-destruction states, described in this book [1–147], will be considered as a close to the representative open macroscopic quantum system (OMQS) under the specified conditions. Hence, OMQS are quantum subsystems, i.e., open quantum systems that are in inevitable permanent interaction with other physical systems, which may be named environment. Could the theory of electroviscoelasticity, here presented, be useful in discussion and/or further elucidation related to the problems of the experimental and theoretical status of decoherence? This question is one important subject of the present and future related research!

Such an approach and consequent knowledge could be implemented in the studies of, for example, ionics, spintronics, fractional quantum Hall effect fluids, decoherence sensitivity, quantum computation, entities/quantum particles entanglement, open macroscopic quantum systems, and macroscopic quantum tunneling [1, 11, 137–147].

5.1.3 Concluding Remarks

Up to now, three possible mathematical formalisms have been developed and discussed related to the presented theory of electroviscoelasticity. The first is the stretching tensor model, where the normal and tangential forces are considered regardless (only from a mathematical point of view) of their origin (mechanical and/or electrical). The second is the van der Pol integral-derivative model. Finally, the third model presents an effort to generalize the previous van der Pol integral-differential equations, both linear and non-linear, where the ordinary time derivatives and integrals are replaced by corresponding fractional-order time derivatives and integrals of the order $p < 2$ ($p = n-\delta$, $n = 1, 2, \delta \ll 1$).

Each of these mathematical formalisms, although related to the same physical formalism, facilitates better understanding of different aspects of a droplet existence (formation, life, and destruction).

The stretching tensor model discusses the force equilibrium at the interfaces, either deformable or rigid, but its solution is difficult because the tensor contains nonlinear and complex elements.

The van der Pol derivative model is convenient for discussion of the *antenna output circuit*, the resulting equivalent electrical circuit; but, since in the case of nonlinear oscillators, that is here the realistic one, the problem of determining the noise output is complicated by the fact that the output is fed back into the system, thus modifying in a complicated manner the effective noise input. The noise output appears as an induced anisotropic effect.

The theory of electroviscoelasticity using a generalization of the van der Pol derivative model applying a fractional approach constitutes a new interdisciplinary tool for colloid and interface science. Hence, more degrees of freedom are in the model, memory storage considerations and hereditary properties are included in the model, and histories or impacts to the present and future are in the game!

One may note that the fractional differential operator is not a local operator, that is, the derivative is not only dependent on the value at the point but also the value of the function on the whole interval.

6

Electrophoresis

6.1 Introduction

Referring to Figure 6.1, a charged colloidal particle suspended in an electrolyte solution is surrounded by a diffuse cloud of ions containing a net amount of charge which is roughly equal but opposite in sign to that of the particle. The linear size of this ionic cloud, $1/\kappa$, known as the Debye length or the thickness of the electrical double layer surrounding the particle, depends mainly upon the bulk concentration of electrolytes. For instance, if the liquid phase contains N types of ionic species with bulk number concentration n_{j0}, $j = 1,2,...,N$, then $\kappa = [\sum_{j=1}^{N} n_{j0}(ez_j)^2/\varepsilon k_B T]^{1/2}$, where z_j, e, ε, k_B, and T are the valence of ionic species j, the elementary charge, the permittivity of the liquid phase, the Boltzmann constant, and the absolute temperature, respectively. When an external electric field is applied to the system shown in Figure 6.1, the positively charged particle migrates toward a cathode and the ions in the double layer toward an anode. The phenomenon that a particle is driven by an applied electrical field is called *electrophoresis*.

Being one of the electrokinetic phenomena—the others are electroosmosis, sedimentation potential, and streaming potential—electrophoresis has been studied extensively in the last few decades. This arises from the fact that electrophoresis has been adopted widely in both laboratory- and industrial-scaled separation-purification processes. Electrophoresis can also be applied as a convenient and efficient tool to characterize the surface properties of particles of colloidal size for both inorganic and biological entities. Modern techniques such as electrophoresis display, protein separation, and DNA analysis, to name a few, all involve this phenomenon.

Although the principle of electrophoresis is straightforward, the design of an electrophoresis device can be changeable. This is because the equations governing this phenomenon, known as *electrokinetic equations*, including those for the electrical potential field, the concentration file, and the flow field, are coupled, nonlinear differential equations, and therefore solving them simultaneously is nontrivial, even for the simplest case of an isolated particle in an infinite liquid medium. The earliest available literature of electrokinetic phenomena is that of Reuss [1], who observed the electrophoresis

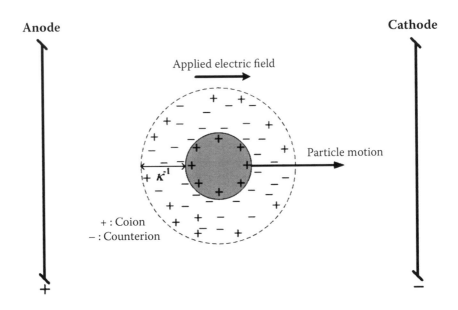

FIGURE 6.1
Electrophoresis of a positively charged colloidal particle subject to an applied external electric field.

of clay particles and the electroosmotic flow of water through a bed of quartz sand. Quincke [2] observed the electroosmotic flow in a glass capillary through a microscope. It was suggested that the electric potential difference across a membrane comes from streaming potential. The development of the electrical double-layer theory was originated by Helmholtz in 1879 [3], when the electrokinetic phenomena were described by applying an electric field across two parallel glass plates. Helmholtz observed a bulk liquid flow toward the negative electrode and concluded that the plate surface acquired negative charge while the solvent close to the surface was positively charged. The description of Helmholtz was refined by Smoluchowski [4], who considered the electrophoresis of a spherical, nonconducting particle of constant surface potential in an infinite electrolyte solution. Assuming symmetric electrolytes; an infinitely thin double layer, $\kappa a \to \infty$, where a is the linear size of the particle; and low surface potential, $eZ\zeta_p/k_BT < 1$, where ζ_p is the surface (zeta) potential of the particle and Z is the valence of electrolytes, he was able to derive the expression below:

$$\mu = \frac{U}{E_\infty} = \frac{\varepsilon\zeta_p}{\eta} \tag{6.1}$$

where U, μ, E_∞, ε, and η are the electrophoretic velocity and the electrophoretic mobility of the particle, the strength of the applied electric field, and the dielectric constant and the viscosity of the electrolyte solution, respectively. Although it is derived under limiting conditions, Eq. (6.1), which is usually referred to as the *Helmholtz-Smoluchowski equation*, or simply *Smoluchowski's equation*, is widely applied in the design of electrophoresis devices for its conciseness and parsimoniousness, and to interpret experimental data. Note that Eq. (6.1) is not limited to spherical particles. According to Eq. (6.1), the electrophoretic mobility of a particle, defined as its velocity per unit strength of the applied electric field, is proportional to the surface potential of the particle, and inversely proportional to the viscosity of the liquid medium. As will be shown in this chapter, these are realistic because the surface potential of an isolated particle is proportional to its surface charge density, and, therefore, the higher the surface potential the greater the electrical driving force acting on the particle, and the larger the viscosity the greater the drag on the particle. On the other extreme, where the double layer surrounding a particle is infinitely thick, that is, $\kappa a \to 0$, the following can be shown [5]:

$$\mu = \frac{2}{3} \frac{\varepsilon \zeta_p}{\eta} \tag{6.2}$$

This expression is known as *Hückel's equation*, which is applicable to large Debye lengths, in particular to non-electrolyte systems such as organic liquid medium. For intermediate values of κa, where the thickness of the double layer surrounding a particle takes a finite value, Henry [6] derived the following expressions for spherical particles:

$$\mu = \frac{\varepsilon \zeta_p}{\eta} f(\kappa a) \tag{6.3}$$

$$f(\kappa a) = \frac{2}{3} + \frac{1}{24}(\kappa a)^2 - \frac{5}{72}(\kappa a)^3 - \frac{1}{144}(\kappa a)^4 + \frac{1}{144}(\kappa a)^5$$
$$+ \frac{1}{12}(\kappa a)^4 e^{\kappa a} \left(1 - \frac{(\kappa a)^2}{12} \right) E_1(\kappa a) \tag{6.4}$$

where $E_n(\kappa a)$ is the exponential integral of order n defined by

$$E_n(\kappa a) = (\kappa a)^{n-1} \int_{\kappa a}^{\infty} \frac{e^{-t}}{t^n} dt \tag{6.5}$$

The derivation of Eqs. (6.3)–(6.5) is based on two main assumptions of Henry. Firstly, the electric double layer surrounding the particle is not distorted by the flow field, and the total potential within the double layer is the sum of the equilibrium potential (the potential in the absence of the applied electric field) and the perturbed potential (the potential arising from the applied electric field). Secondly, the surface potential of the particle is low,

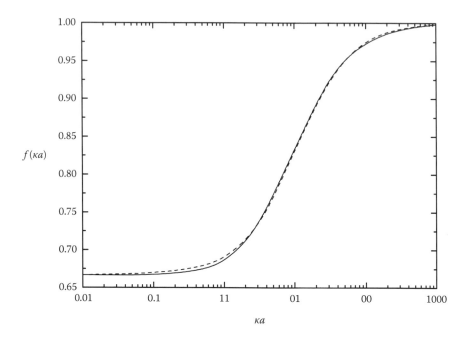

FIGURE 6.2
Variation of Henry's function as a function of κa. Solid line, Eq. (6.4); dashed line, Eq. (6.6).

that is, the Debye-Hückel condition is applicable. Because it is applicable to $0 < \kappa a < \infty$, which covers essentially all the conditions in practice, Henry's expression is widely adopted. The function $f(\kappa a)$ in Eq. (6.3) is usually called *Henry's function*. The solid curve in Figure 6.2 shows the behavior of $f(\kappa a)$.

Figure 6.2 suggests that $f(\kappa a) \to 2/3$ as $\kappa a \to 0$, and $f(ka) \to 1$ as $\kappa a \to \infty$. That is, both Hückel's equation (Eq. [6.2]) and the Helmholtz-Smoluchowski equation (Eq. [6.1]) can be recovered as the limiting cases of Henry's equation, Eq. (6.3). Table 6.1 illustrates the values of Henry's function $f(\kappa a)$ calculated numerically by Eq. (6.4).

Based on a curve-fitting procedure, the following expression can be obtained [7]:

$$f(\kappa a) = 1 - \frac{1}{3[1 + a_1(\kappa a)^{a_2}]} \tag{6.6}$$

where $a_1 = 0.07234$ and $a_2 = 1.129$. As seen in Figure 6.2, this performance of this empirical expression is satisfactory.

Because it is valid for arbitrary double-layer thickness, Henry's formula, Eq. (6.3), provides a very useful and convenient result for predicting the electrophoretic mobility and/or interpreting the experimentally observed electrophoretic

TABLE 6.1

Values of Henry's Function
Evaluated Numerically by Eq. (6.4)

κa	$f(\kappa a)$
0	0.6667
0.5	0.6727
1.0	0.6847
2.0	0.7100
3.0	0.7340
4.0	0.7553
5.0	0.7733
10.0	0.8333
25.0	0.9133
50.0	0.9533
100.0	0.9733
∞	1

behavior of a particle. However, it is limited to the case of low surface potential and an isolated spherical particle. The former, also known as the *Debye-Hückel condition*, is satisfied if the surface potential is lower than circa 25 mV, which can be violated in many cases of practical significance. In the latter, the influences of the presence of a boundary and neighboring particles are neglected. The boundary effect can be significant, for example, if electrophoresis is conducted in a narrow space such as capillary electrophoresis where the wall of a device should not be neglected. The influence of neighboring particles can be important in a case, for example, in which the concentration of particles is appreciable. Another factor of practical significance neglected in Henry's formula is double-layer polarization, which arises from the convective motion of ionic species. This effect was analyzed first by Overbeek [8] and Booth [9] by considering the electrophoresis of a nonconducting spherical particle under the conditions of low surface potential. This analysis was extended by Wiersema et al. [10] to the case of high surface potentials through solving numerically the governing equations for the case of binary electrolytes, taking account of the effect of ionic convection. The mobility of a particle was tabulated for various surface potentials. However, for a 1:1 electrolyte solution, if the surface potential exceeds circa 150 mV, or circa 25 mV for a 2:2 electrolyte solution, their iterative numerical scheme might not converge. The effect of double-layer polarization was also modeled by O'Brien and White [11] by considering arbitrary levels of surface potential and the thickness of the double layer. The governing equations were solved numerically by adopting the integration scheme of Nordsieck [12]. All these studies are limited to an isolated particle in an infinite fluid medium. Because the concentration, the electric, and the flow fields are all affected, the presence of a boundary can have a profound influence on the electrophoretic behavior of a particle. Since the degree of difficulty in solving the electrokinetic equations depends largely upon the level of the electrical potential, the geometry of the problem, and the types of boundary conditions, the presence of a boundary and/or neighboring particles can make the resolution procedure complicated.

Extending the classic electrophoresis theory so that it is capable of reflecting more realistically the conditions of practical significance is highly desirable to experimentalists. To this end, a considerable amount of effort has been made in recent decades. Regarding the boundary effect, for instance, the modification of Smoluchowski's analysis [4] was begun by assuming an infinitely thin double layer [13–24], and the specific effect of the double layer is incorporated into the hydrodynamic equations through assuming a slip boundary condition. In these studies a particle is driven by a surface stress that comes from the slip velocity boundary condition, which is appropriate for rigid particles under typical conditions but becomes unrealistic for soft or fuzzy particles such as polyelectrolytes and biocolloids where their surfaces are penetrable to ionic species. Subsequent analyses, which are capable of modeling the case of an arbitrary thick double layer, were made in many studies by considering several types of boundary effects [25–65].

This chapter is aimed at providing a brief overview of electrophoresis. The introduction has begun by presenting the general electrokinetic equations and the associated boundary conditions for the case of an isolated charged particle in an infinite electrolyte solution. A perturbation approach [11, 66], which is based on the condition of a weak applied electric field, is adopted to partition the general electrokinetic equations into an equilibrium component and a perturbed component so that the general behavior of the system under consideration can be recovered. The boundary effect on electrophoresis is discussed briefly, followed by the introduction of the boundary conditions. The procedure for the evaluation of the electrical force and the hydrodynamic force acting on a particle is then introduced. The former is based on the Maxwell stress tensor, and the latter on the hydrodynamic stress tensor [66]. Using the electrophoresis of a spherical particle in a spherical cavity as an illustrating example, the solution procedure, which includes both analytical [4–6,25] and numerical [27,44,55,66] methods, is presented. For illustration, the particles considered are rigid and nonconducting, and the liquid phase is an incompressible Newtonian fluid.

6.2 General Governing Equations

Referring to Figure 6.3, let us consider first the electrophoresis of an isolated, rigid, nonconductive, charged spherical particle with a radius of a in an infinite electrolyte solution medium containing N kinds of ionic species of valence z_j and bulk concentration n_{j0}. A uniform electric field $\mathbf{E}_\infty$ of strength

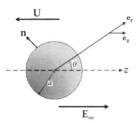

FIGURE 6.3

A sphere of radius a in an infinite fluid medium moves with velocity **U** as a response to an applied uniform electric field **E**$_\infty$ parallel to the z direction.

Note: **n** is the unit's outer normal vector on the particle surface; **e**$_r$ and **e**$_z$ are the unit vectors in the r and z directions, respectively.

E_∞ is applied in the z direction. Let U be the electrophoretic velocity of the particle along the z direction. Suppose that the surface of the particle is non-slip. For convenience, we let the particle be fixed, and the liquid far away from the particle moves with a relative velocity $\mathbf{u}_\infty = -U\mathbf{e}_z$, with $\mathbf{e}_z$ being the unit vector in the z direction.

The governing equations of the present problem include those for the electrical, the flow, and the concentration fields. If we let ϕ be the electrical potential, then it satisfied the Poisson equation

$$\nabla^2\phi = -\frac{\rho_e}{\varepsilon} = -\sum_{j=1}^{N}\frac{z_j e n_j}{\varepsilon}, \qquad (6.7)$$

where ∇^2 is the Laplace operator, ε is the permittivity of the liquid phase, ρ_e is the space charge density, e is the elementary charge, and n_j and z_j are the number concentration and the valence of ionic species j, respectively.

The Reynolds number in electrophoresis is on the order of 10^{-4}, that is, the fluid flow is in the creeping flow regime. Therefore, the flow field at steady state can be described by the continuity equation and the modified Navier-Stokes equation:

$$\nabla\cdot\mathbf{u} = 0 \qquad (6.8)$$

$$\eta\nabla^2\mathbf{u} - \nabla P - \rho_e\nabla\phi = 0 \qquad (6.9)$$

In these expressions, ∇ is the gradient operator; $\mathbf{u}$, η, and P are the velocity, the viscosity, and the pressure of the liquid phase, respectively; and $-\rho_e\nabla\phi$ is the electric body force acting on the liquid.

In the absence of any chemical reactions in the liquid phase, the conservation of each ionic species yields

$$\nabla\cdot\mathbf{f}_j = 0, \quad j = 1, 2, \ldots, N, \qquad (6.10)$$

where $\mathbf{f}_j$ is the flux density of ionic species j, which can be described by the Nernst-Planck equation:

$$\mathbf{f}_j = n_j \mathbf{v}_j = n_j \mathbf{u} - D_j \left(\nabla n_j + \frac{z_j e n_j}{k_B T} \nabla \phi \right), \quad j = 1, 2, \dots, N \qquad (6.11)$$

Here, n_j, D_j, and $\mathbf{v}_j$, are the number concentration, the diffusion coefficient, and the velocity of ionic species j, respectively; e is the elementary charge; k_B is the Boltzmann constant; and T is the absolute temperature.

6.3 Boundary Conditions

The boundary conditions associated with the electric, the flow, and the concentration fields on the particle surface include the following. Because the permittivity of the rigid particle is much smaller than that of the liquid phase, assuming Gauss's law leads to the following expression on the particle surface:

$$\mathbf{n} \cdot \nabla \phi = -\frac{\sigma_p}{\varepsilon} \qquad (6.12)$$

where $\mathbf{n}$ is the unit outer normal vector on the particle surface, and σ_p is the surface charge density of the particle. The no-slip nature of the particle surface yields

$$\mathbf{u} = 0 \qquad (6.13)$$

If the particle surface is impermeable to ionic species, then

$$\mathbf{n} \cdot \mathbf{f}_j = 0 \qquad (6.14)$$

For the present case, the other boundary is at a point far away from the particle, where the following conditions are assumed:

$$\nabla \phi = -\mathbf{E}_\infty \qquad (6.15)$$

$$\mathbf{u}_\infty = -U \mathbf{e}_z \qquad (6.16)$$

$$n_j = n_{j0} \qquad (6.17)$$

These expressions state that at a point far away from the particle, the electric, the flow, and the concentration fields are not influenced by the particle.

To evaluate the electrophoretic velocity of the particle, Eqs. (6.7)–(6.11) need to be solved simultaneously subject to the conditions expressed in Eqs. (6.12)–(6.17). Under general conditions, solving this problem analytically is almost impossible due to the coupled, nonlinear nature of the partial differential equations involved. In practice, this difficulty is circumvented either by deriving approximate analytical results or by resorting to numerical methods. The former is usually based on a perturbation approach [11, 66–68], which is applicable to an arbitrary level of surface potential, the thickness of the double layer, the geometry of a problem, and the types of boundary conditions. For illustration, we focus on this approach in subsequent discussions.

6.4 Perturbation Approach

Suppose that the applied electric field is relatively weak compared with that established by the charged particle. In practice, the surface potential of a particle is lower than circa 250 mV, and the Debye length ranges from 10 nm to 1 µm. This implies that the strength of the electric field established by the particle ranges from 2.5×10^5 to 2.5×10^7 kV/m. Because the strength of the applied electric field is much lower than this level, the assumption of a weak applied electric field is realistic. In this case, $\mathbf{u}$, P, ϕ, n_j, and ρ_e can be partitioned into an equilibrium term, denoted with a superscript (e), and a perturbed term, denoted with a prefix δ, as [11, 66–68] follows:

$$\mathbf{u} = \mathbf{u}^{(e)} + \delta\mathbf{u} \tag{6.18}$$

$$P = p^{(e)} + \delta p \tag{6.19}$$

$$\phi = \phi^{(e)} + \delta\phi \tag{6.20}$$

$$n_j = n_j^{(e)} + \delta n_j \tag{6.21}$$

$$\rho_e = \rho_e^{(e)} + \delta\rho_e \tag{6.22}$$

The equilibrium terms arise from the presence of the particle in the absence of the applied electric field $\mathbf{E}_\infty$, and the perturbed terms arise from the application of $\mathbf{E}_\infty$. Note that because the particle should be stagnant if $\mathbf{E}_\infty$ is not applied, $\mathbf{u}^{(e)} = 0$. Substituting Eqs. (6.18)–(6.22) into the general governing equations, Eqs. (6.7)–(6.11), yields an equilibrium problem and a perturbed problem.

6.4.1 Equilibrium Problem

Collecting terms involving equilibrium quantities, we obtain the following set of equations for the equilibrium electrical, flow, and concentration fields:

$$\nabla^2 \phi^{(e)} = -\frac{\rho_e^{(e)}}{\varepsilon} = -\sum_{j=1}^{N} \frac{z_j e n_j^{(e)}}{\varepsilon} \qquad (6.23)$$

$$-\nabla p^{(e)} + \varepsilon \nabla^2 \phi^{(e)} \nabla \phi^{(e)} = 0 \qquad (6.24)$$

$$\nabla n_j^{(e)} + \frac{z_j e n_j^{(e)}}{k_B T} \nabla \phi^{(e)} = 0 \qquad (6.25)$$

Integrating Eq. (6.25) from a point far away from the particle, where $\phi^{(e)} = 0$ and $n_j^{(e)} = n_{j0}$, to a point near the particle, where $\phi^{(e)} = \phi^{(e)}$ and $n_j^{(e)} = n_j^{(e)}$, leads to the Boltzmann distribution:

$$n_j^{(e)} = n_{j0} \exp\left(-\frac{z_j e \phi^{(e)}}{k_B T}\right), \quad j = 1, 2, \dots, N \qquad (6.26)$$

Substituting Eq. (6.26) into Eq. (6.23) gives

$$\nabla^2 \phi^{(e)} = -\sum_{j=1}^{N} \frac{z_j e n_{j0}}{\varepsilon} \exp\left(-\frac{z_j e \phi^{(e)}}{k_B T}\right) \qquad (6.27)$$

This is known as the *Poisson-Boltzmann equation*, which describes the electric potential distribution inside a stationary double layer. As pointed out by Hunter [69], although it is not exact on strictly statistical mechanical grounds, the Poisson-Boltzmann equation is sufficiently accurate from the applications point of view. Equations (6.23) and (6.24) lead to

$$-\nabla p^{(e)} = \varepsilon \sum_{j=1}^{N} \frac{z_j e n_{j0}}{\varepsilon} \nabla \phi^{(e)} \qquad (6.28)$$

Integrating this expression from $p^{(e)} = p_\infty$ and $n_j^{(e)} = n_{j0}$ to $p^{(e)} = p^{(e)}$ and $n_j^{(e)} = n_j^{(e)}$, we obtain

$$p^{(e)} = p_\infty + \sum_{j=1}^{N} k_B T \left(n_j^{(e)} - n_{j0}\right) \qquad (6.29)$$

where p_∞ is the equilibrium pressure of the fluid at a point far from the particle. This expression describes the equilibrium or static pressure distribution.

6.4.2 Perturbed Problem

The polarization of the double layer surrounding a particle is known to have a significant effect on its electrophoretic behavior [10,11]. In general, because the polarized double layer yields a local electric field, the direction of which is opposite that of the applied electric field, therefore the electrophoretic velocity of the particle is slower than is the case when the double layer is not polarized. Under the condition of a weak applied electric field, O'Brien and White [11] proposed using the following expression to take the effect of double-layer polarization into account:

$$n_j = n_{j0}\exp\left(-\frac{z_j e\left(\phi^{(e)} + \delta\phi + g_j\right)}{k_B T}\right), \quad j = 1, 2, \ldots, N \tag{6.30}$$

Here, g_j is a hypothesized potential function, which is the key for the description of a polarized double layer. Note that although Eq. (6.30) seems to be of the Boltzmann type, the actual distribution of n_j depends upon g_j.

Equations (6.27) and (6.30) and $\phi = \phi^{(e)} + \delta\phi$ yield the governing equation for the perturbed potential $\delta\phi$:

$$\nabla^2\delta\phi = \nabla^2\phi - \nabla^2\phi^{(e)}$$

$$= -\sum_{j=1}^{N} \frac{z_j e n_{j0}}{\varepsilon}\left[\exp\left(-\frac{z_j e\left(\phi^{(e)} + \delta\phi + g_j\right)}{k_B T}\right) - \exp\left(-\frac{z_j e\phi^{(e)}}{k_B T}\right)\right] \tag{6.31}$$

Equations (6.21), (6.26), and (6.30) lead to

$$\delta n_j = n_{j0}\exp\left(-\frac{z_j e\left(\phi^{(e)} + \delta\phi + g_j\right)}{k_B T}\right) - n_{j0}\exp\left(-\frac{z_j e\phi^{(e)}}{k_B T}\right), \quad j = 1, 2, \ldots, N \tag{6.32}$$

Substituting Eqs. (6.18), (6.20), and (6.32) into Eqs. (6.10) and (6.11), we obtain

$$\nabla^2 g_j - \frac{z_j e}{k_B T}\nabla\phi^{(e)} \cdot \nabla g_j$$

$$= \frac{1}{D_j}\delta\mathbf{u} \cdot \nabla\phi + \frac{1}{D_j}\delta\mathbf{u} \cdot \nabla g_j + \frac{z_j e}{k_B T}\nabla\delta\phi \cdot \nabla g_j + \frac{z_j e}{k_B T}\nabla g_j \cdot \nabla g_j, \quad j = 1, 2, \ldots, N \tag{6.33}$$

Suppose that the applied electric field is weak compared to that established by the particle and/or the boundary. In this case, because $\phi^{(e)} \gg \delta\phi$

and $\phi^{(e)} \gg \delta\phi + g_j$ [11], expanding Eq. (6.30) into the Taylor series and neglecting higher-order terms, we obtain

$$n_j = n_{j0}\left[\exp\left(-\frac{z_j e\phi^{(e)}}{k_B T}\right)\right]\left[1 - \frac{z_j e}{k_B T}(\delta\phi + g_j)\right], \quad j = 1, 2, \dots, N \qquad (6.34)$$

This expression describes the concentration of ionic species j after the external electric field is applied. Because $\delta\mathbf{u}$, δp, $\delta\phi$, and δn_j are all on the order of E_∞, by substituting Eq. (6.34) into Eqs. (6.31) and (6.33) and neglecting terms involving two or more of those perturbed terms, it can be shown that the disturbed electric fields can be approximated by

$$\nabla^2\delta\phi = -\sum_{j=1}^{N}\frac{z_j e n_{j0}}{\varepsilon}\left[-\exp\left(-\frac{z_j e\phi^{(e)}}{k_B T}\right)\right]\left[-\frac{z_j e}{k_B T}(\delta\phi + g_j)\right] \qquad (6.35)$$

$$\nabla^2 g_j - \frac{z_j e}{k_B T}\nabla\phi^{(e)}\cdot\nabla g_j = \frac{1}{D_j}\delta\mathbf{u}\cdot\nabla\phi^{(e)}, \quad j = 1, 2, \dots, N \qquad (6.36)$$

For the flow field, substituting Eqs. (6.18)–(6.20) and (6.22) into Eqs. (6.8) and (6.9) and collecting terms of the order E_∞, we obtain

$$\nabla\cdot\delta\mathbf{u} = 0 \qquad (6.37)$$

$$\eta\nabla^2\delta\mathbf{u} - \nabla\delta p + \varepsilon\nabla^2\phi^{(e)}\nabla\delta\phi + \varepsilon\nabla^2\delta\phi\nabla\phi^{(e)} = 0 \qquad (6.38)$$

The linearized expressions, Eqs. (6.27) and (6.35)–(6.38), are applicable to arbitrary zeta potential, double-layer thickness, and geometry. Note that the sum of the last two terms on the right-hand side of Eq. (6.38), $\varepsilon\nabla^2\phi^{(e)}\nabla\delta\phi + \varepsilon\nabla^2\delta\phi\nabla\phi^{(e)}$, which represents the electric body force acting on the liquid, is different from that in some literature [7,14–25,27,28,30–33,35–46,48–51,57,58,61,65,67], $-\rho_e\nabla\phi = \varepsilon\nabla^2\phi\nabla\phi$. The influence of that force on the electrophoretic mobility of a particle will be discussed in detail in this chapter.

6.5 Boundary Effects

Smoluchowski's formula [4], Eq. (6.1), is based on an isolated particle in an infinite fluid. A considerable amount of works [12–64] have been devoted in recent decades to extend that expression to the case where electrophoresis is conducted in a finite space so that the result is capable of describing more realistically the conditions in practice. Up to now, this is still an issue of fundamental significance in many fields. In capillary electrophoresis and electrophoresis of particles through a porous medium, for instance, the presence of a boundary should not be neglected. The presence of a boundary can have a profound influence on the electrophoretic behavior of a particle, and care must be taken in both the interpretation of the experimental data and the design of an electrophoresis apparatus. In general, solving the governing electrokinetic equations for the case when the boundary effect is significant is more difficult than when it is insignificant. If the boundary is uncharged, it can influence the electrophoretic behavior of a particle hydrodynamically through raising the viscous drag acting on the particle, and electrically through distorting the local electric field near the particle [13–65]. The situation becomes more complicated if the boundary is charged. In this case, three additional effects need to be considered, namely, the presence of an electroosmotic flow [25–29,39,40,53–56,60,62–64], the osmotic pressure [29,47,59], and the charge induced on the particle surface as it gets close to the boundary [53–56]. The electrophoretic behavior of a particle can be influenced significantly, both quantitatively and qualitatively, by these effects. For example, the presence of a charged boundary is capable of reversing the direction of the movement of a particle [25,27,28,47,53,56,59,62,63], and the electrophoretic mobility of a particle can exhibit a local maximum and/or minimum as the thickness of the double layer surrounding the particle and/or the boundary varies [29,39,44,47–49,53–56,57,60,62–64]. These results are of practical importance in cases where, for instance, electrophoresis is adopted as a separation tool. Typical example where a charged boundary can play a role includes the capillary electrophoresis conducted in a fused silica pore. For example, when silica is in contact with an aqueous solution, its surface hydrolyzes to form silanol surface groups. Depending upon the pH of the liquid phase, these groups can be positively charged as $Si-OH_2^+$, neutral as Si-OH, or negatively charged as $Si-O^-$ [70,71]. Another typical example is the electrophoresis conducted in a microchip, where the wall surface is modified by a polymer coating, yielding either a positively or a negatively charged surface [72].

To model the electrophoresis of a particle when a boundary is present, let us consider the general case illustrated in Figure 6.4, where a particle of surface

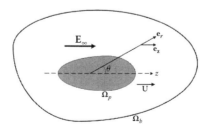

FIGURE 6.4
A colloidal particle of surface Ω_p in a medium confined by a boundary Ω_b.

Note: $\mathbf{E}_\infty$ is a uniform applied electric field parallel to the z axis, and $\mathbf{e}_r$ and $\mathbf{e}_z$ are the unit vectors in the r and z directions, respectively.

Ω_p is placed in a medium bounded by a surface Ω_b. The space between Ω_p and Ω_b is filled with an aqueous solution containing z_1:z_2 electrolytes; z_1 and z_2 are the valences of cations and anions, respectively. A uniform electric field $\mathbf{E}_\infty$ of strength E_∞ is applied in the z direction, and $\mathbf{U}$ is the corresponding electrophoretic velocity of the particle; $\mathbf{e}_r$ and $\mathbf{e}_z$ are the unit vectors in the r and z directions, respectively. Suppose that the liquid phase is an incompressible Newtonian fluid, $\mathbf{E}_\infty$ is relatively weak compared with that established by the particle and/or the boundary, and both Ω_p and Ω_b are nonconductive and nonslip.

6.5.1 Governing Equations

For a more concise presentation, $\phi^{(e)}$, $\delta\phi$, $\delta\mathbf{u}$, and δp in Eqs. (6.27), and (6.34)–(6.38) are replaced respectively by ϕ_1, ϕ_2, $\mathbf{u}$, and p, and these equations are rewritten as follows:

$$\nabla^2\phi_1 = -\sum_{j=1}^{2}\frac{z_j e n_{j0}}{\varepsilon}\exp\left(-\frac{z_j e \phi_1}{k_B T}\right) \tag{6.39}$$

$$\nabla^2\phi_2 = -\sum_{j=1}^{2}\frac{z_j e n_{j0}}{\varepsilon}\left[-\exp\left(\frac{z_j e \phi_1}{k_B T}\right)\right]\left[-\frac{z_j e}{k_B T}(\phi_2 + g_j)\right] \tag{6.40}$$

$$\nabla^2 g_j - \frac{z_j e}{k_B T}\nabla\phi_1 \cdot \nabla g_j = \frac{1}{D_j}\mathbf{u}\cdot\nabla\phi_1, \quad j=1,2 \tag{6.41}$$

$$\nabla \cdot \mathbf{u} = 0 \tag{6.42}$$

$$\eta\nabla^2\mathbf{u} - \nabla p + \varepsilon\nabla^2\phi_1\nabla\phi_2 + \varepsilon\nabla^2\phi_2\nabla\phi_1 = 0 \tag{6.43}$$

$$n_j = n_{j0}\left[\exp\left(-\frac{z_j e \phi_1}{k_B T}\right)\right]\left[1 - \frac{z_j e}{k_B T}(\phi_2 + g_j)\right], \quad j=1,2 \tag{6.44}$$

6.5.2 Low Surface Potential Model

If the surface potential on Ω_p and/or Ω_b is low, then the original problem can be simplified considerably. This assumption is often made in the literature to obtain approximate results. Under that condition, the effect of double-layer polarization can be neglected, $\delta n_j = 0$ [6], and Eqs. (6.39)–(6.43) become as follows [26,29,34,47,52–54]:

$$\nabla^2\phi_1 = \kappa^2\phi_1 \tag{6.45}$$

$$\nabla^2\phi_2 = 0 \tag{6.46}$$

$$\nabla \cdot \mathbf{u} = 0 \tag{6.47}$$

$$\eta \nabla^2 \mathbf{u} - \nabla p + \varepsilon \nabla^2 \phi_1 \nabla \phi_2 = 0 \tag{6.48}$$

In these expressions, $\kappa = [\sum_{j=1}^{2} n_{j0}(ez_j)^2/\varepsilon k_B T]^{1/2}$ is the reciprocal Debye length. The last term on the right-hand side of Eq. (6.48) denotes the electric body force, where $-\varepsilon\nabla^2\phi_1 = \rho_e$ is the space charge density, and $-\nabla\phi_2 = \mathbf{E}_\infty$ is the applied electric field [26,29,34,47,52–54]. In the literature, the electric body force is represented by $-\rho_e\nabla\phi = \varepsilon\nabla^2\phi\nabla\phi$ in many studies [7,14–25,27, 28,30–33,35–46,48–51,57,58,61,65,67], that is, Eq. (6.43) or Eq. (6.48) is replaced by

$$\eta \nabla^2 \mathbf{u} - \nabla p + \varepsilon \nabla^2 \phi \nabla \phi = 0 \tag{6.49}$$

Note that for a system of not totally symmetric nature, such as a particle moving normal to a planar surface [14–21,31,32,46], a particle at an arbitrary position in a spherical cavity [35,37,42], and other types of geometry [23,39,43,48–51], using Eq. (6.49) could be inappropriate. This is because an extraneous electric force, $\varepsilon\nabla^2\phi_1\nabla\phi_1$, is included in the evaluation of the electrophoretic velocity of a particle, implying that the particle is also driven by the electric field established by the equilibrium electric potential, $-\nabla\phi_1$, and, therefore, its mobility will be overestimated [52,53,73].

6.5.3 Boundary Conditions

Let us introduce next the boundary conditions associated with the governing equation for the electric and the flow fields for rigid particles and boundaries.

6.5.3.1 Electric Field

Three types of boundary conditions are usually considered in the literature for the electric field, namely, the constant surface potential [11,13–30,32,35, 37–40,43–46,48–56,58,60,63–65,73], the constant surface charge density [34,35, 52,60], and charge-regulated surfaces [31,36,42,47,59,62]. Mathematically, the first two types of boundary condition can be recovered from the last type of boundary condition as limiting cases, and the behavior of a particle in the last case is bounded by the corresponding behaviors of the particle in the first two cases.

If Ω_p and Ω_b are maintained at potentials ζ_p and ζ_b, respectively, then the boundary conditions associated with the equilibrium potential are

$$\phi_1 = \zeta_p \quad \text{on} \quad \Omega_p \tag{6.50}$$

$$\phi_1 = \zeta_b \quad \text{on} \quad \Omega_b \tag{6.51}$$

For the case in which Ω_p is maintained at a constant surface charge density, σ_p, then applying Gauss's law around Ω_p yields

$$(\mathbf{n} \cdot \varepsilon \nabla \phi_1)_{\Omega_p^+} - (\mathbf{n} \cdot \varepsilon \nabla \phi_1)_{\Omega_p^-} = -\sigma_p \tag{6.52}$$

where $\mathbf{n}$ is the unit normal vector directed into the liquid phase, and Ω_p^- and Ω_p^+ denote a boundary immediately inside and outside Ω_p, respectively. Because the permittivity of the rigid particle is much smaller than that of the liquid phase, Eq. (6.52) can be approximated by

$$\mathbf{n} \cdot \nabla \phi_1 = \frac{-\sigma_p}{\varepsilon} \quad \text{on} \quad \Omega_p \tag{6.53}$$

Similarly, if the boundary is a rigid surface which is maintained at a constant charge density σ_b, then

$$\mathbf{n} \cdot \nabla \phi_1 = \frac{-\sigma_b}{\varepsilon} \quad \text{on} \quad \Omega_b \tag{6.54}$$

The constant surface potential and the constant surface charge density are special cases of a more general type of boundary condition. Mathematically, the boundary conditions expressed in Eqs. (6.50) and (6.51) are of the Dirichlet type, and those in Eqs. (6.53) and (6.54) are of the Neumann type. It is possible that the boundary conditions are a combination of these two types, that is, a mixed or Robin type. This occurs, for example, in the case of a charge-regulated surface. Let us consider, for example, the following dissociation or association reactions occurring on the surface of a particle [74]:

$$AH \leftrightarrow A^- + H^+ \tag{6.55}$$

$$BH^+ \leftrightarrow B + H^+ \tag{6.56}$$

If we let K_a and K_b be the equilibrium constants of these reactions, then

$$K_a = \frac{[A^-]_{\Omega_p}[H^+]_{\Omega_p}}{[AH]_{\Omega_p}} \tag{6.57}$$

$$K_b = \frac{[B]_{\Omega_p}[H^+]_{\Omega_p}}{[BH^+]_{\Omega_p}} \tag{6.58}$$

Here, symbols with subscript Ω_p denote properties on the particle surface. $[A^-]_{\Omega_p}$, $[AH]_{\Omega_p}$, $[B]_{\Omega_p}$, $[BH^+]_{\Omega_p}$, and $[H^+]_{\Omega_p}$ are the numbers of A^-, AH,

B, BH$^+$, and H$^+$ per unit area of Ω_p, respectively. If we let $N_{\Omega_p,A}$ and $N_{\Omega_p,B}$ be the total number of acidic and basic functional groups per unit area of Ω_p, respectively, then

$$N_{\Omega_p,A} = [A^-]_{\Omega_p} + [AH]_{\Omega_p} \tag{6.59}$$

$$N_{\Omega_p,B} = [BH^+]_{\Omega_p} + [B]_{\Omega_p} \tag{6.60}$$

Suppose that the spatial concentration of H$^+$ follows the Boltzmann distribution [69,75]. Then

$$[H^+]_{\Omega_p} = [H^+]_b \exp\left(\frac{-ez_1\zeta_p}{k_BT}\right) \tag{6.61}$$

where $[H^+]_b$ is the bulk concentration of H$^+$. Equations (6.58), (6.60), and (6.61) yield

$$[BH^+]_{\Omega_p} = \frac{N_{\Omega_p,B}\left(\frac{[H^+]_b}{K_b}\right)\exp\left(-\frac{ez_1\zeta_p}{k_BT}\right)}{1+\left(\frac{[H^+]_b}{K_b}\right)\exp\left(-\frac{ez_1\zeta_p}{k_BT}\right)} \tag{6.62}$$

Therefore, the charge density of BH$^+$ on Ω_p, $\sigma_{p,B} = e[BH^+]_{\Omega_p}$, is

$$\sigma_{p,B} = \frac{eN_{\Omega_p,B}\left(\frac{[H^+]_b}{K_b}\right)\exp\left(-\frac{ez_1\zeta_p}{k_BT}\right)}{1+\left(\frac{[H^+]_b}{K_b}\right)\exp\left(-\frac{ez_1\zeta_p}{k_BT}\right)} \tag{6.63}$$

Similarly, combining Eqs. (6.57), (6.59), and (6.61) into $\sigma_{p,A} = -e[A^-]_{\Omega_p}$, the charge density of A$^-$ on Ω_p, $\sigma_{p,A}$, can be expressed as

$$\sigma_{p,A} = -\frac{eN_{\Omega_p,A}}{1+\left(\frac{[H^+]_b}{K_a}\right)\exp\left(-\frac{ez_1\zeta_p}{k_BT}\right)} \tag{6.64}$$

and the boundary condition associated with ϕ_1 on the particle surface becomes

$$\mathbf{n}\cdot\nabla\phi_1 = -\frac{\sigma_{p,A}+\sigma_{p,B}}{\varepsilon}$$

$$= \frac{\left(\frac{eN_{\Omega_p,A}}{\varepsilon}\right)}{1+\left(\frac{[H^+]_b}{K_a}\right)\exp\left(-\frac{ez_1\zeta_p}{k_BT}\right)} - \frac{\left(\frac{eN_{\Omega_p,B}}{\varepsilon}\right)\left(\frac{[H^+]_b}{K_b}\right)\exp\left(-\frac{ez_1\zeta_p}{k_BT}\right)}{1+\left(\frac{[H^+]_b}{K_b}\right)\exp\left(-\frac{ez_1\zeta_p}{k_BT}\right)} \tag{6.65}$$

where $\sigma_{p,A} + \sigma_{p,B} = \sigma_p$. In terms of scaled symbols, this expression becomes

$$\mathbf{n} \cdot \nabla^* \phi_1^* = \frac{A}{1 + B\exp(-\phi_1^*)} - \frac{\Phi A B \exp(-\phi_1^*)}{\Omega[1 + \Phi B \exp(-\phi_1^*)]} \tag{6.66}$$

where $\nabla^* = a\nabla$, $\phi_1^* = \phi_1/(k_B T/e z_1)$, $A = e^2 z_1 N_{\Omega_p,A} a/\varepsilon k_B T$, $B = [H^+]_b/K_a$, $\Omega = N_{\Omega_p,A}/N_{\Omega_p,B}$, and $\Phi = K_a/K_b$ [59], and a is the characteristic length of a particle (the radius for spherical particles) [28]. If ζ_p is less than circa 25 mV, $\sigma_{p,B}$ can be approximated by the following [59]:

$$\sigma_{p,B} = \sigma_{p,B}\big|_{\zeta_p=0} + \frac{d\sigma_{p,B}}{d\zeta_p}\bigg|_{\zeta_p=0} (\zeta_p - 0)$$

$$= \frac{eN_{\Omega_p,B}\left(\frac{[H^+]_b}{K_b}\right)}{1 + \left(\frac{[H^+]_b}{K_b}\right)} - \frac{\left(\frac{e^2 z_1 N_{\Omega_p,B}}{k_B T}\right)\left(\frac{[H^+]_b}{K_b}\right)}{\left[1 + \left(\frac{[H^+]_b}{K_b}\right)\right]^2} \zeta_p \tag{6.67}$$

Similarly, $\sigma_{p,A}$ can be expressed as follows [47,59]:

$$\sigma_{p,A} = \frac{-eN_{\Omega_p,A}}{1 + \left(\frac{[H^+]_b}{K_a}\right)} - \frac{\left(\frac{e^2 z_1 N_{\Omega_p,A}}{k_B T}\right)\left(\frac{[H^+]_b}{K_a}\right)}{\left[1 + \left(\frac{[H^+]_b}{K_a}\right)\right]^2} \zeta_p \tag{6.68}$$

Therefore, if the surface potential is low, the boundary condition associated with ϕ_1 on the particle surface becomes

$$\mathbf{n} \cdot \nabla \phi_1 = \frac{\left(\frac{eN_{\Omega_p,A}}{\varepsilon}\right)}{1 + \left(\frac{[H^+]_b}{K_a}\right)} + \frac{\left(\frac{e^2 z_1 N_{\Omega_p,A}}{\varepsilon k_B T}\right)\left(\frac{[H^+]_b}{K_a}\right)}{\left[1 + \left(\frac{[H^+]_b}{K_a}\right)\right]^2} \phi_1$$

$$- \frac{\left(\frac{eN_{\Omega_p,B}}{\varepsilon}\right)\left(\frac{[H^+]_b}{K_b}\right)}{1 + \left(\frac{[H^+]_b}{K_b}\right)} + \frac{\left(\frac{e^2 z_1 N_{\Omega_p,B}}{\varepsilon k_B T}\right)\left(\frac{[H^+]_b}{K_b}\right)}{\left[1 + \left(\frac{[H^+]_b}{K_b}\right)\right]^2} \phi_1 \tag{6.69}$$

In terms of the scaled symbols, this expression becomes

$$\mathbf{n} \cdot \nabla^* \phi_1^* = \frac{A}{1 + B} + \frac{AB}{(1 + B)^2}\phi_1^* - \frac{\Phi A B}{\Omega(1 + \Phi B)} + \frac{\Phi A B}{\Omega(1 + \Phi B)^2}\phi_1^* \tag{6.70}$$

Note that the constant surface charge density model, Eq. (6.53), can be recovered from the charge-regulated model, Eqs. (6.66) or (6.70), by letting $A \to 0$ and choosing a sufficiently small ($\ll 0{,}01$) or a sufficiently large ($\gg 100$) value of B [47,59]. Similarly, the constant surface potential model, Eq. (6.50) [47], can be recovered by letting $A \to \infty$ in Eq. (6.70).

Under the conditions that both Ω_p and Ω_b are nonconductive and impenetrable to ionic species, and the concentration of ionic species reaches the equilibrium value on the boundary surface, the boundary conditions for ϕ_2 and g_j can be expressed as follows [11,28]:

$$\mathbf{n} \cdot \nabla \phi_2 = 0 \quad \text{on} \quad \Omega_p \tag{6.71}$$

$$\mathbf{n} \cdot \nabla \phi_2 = -E_\infty \cos\theta \quad \text{on} \quad \Omega_b \tag{6.72}$$

$$\mathbf{n} \cdot \nabla g_j = 0 \quad \text{on} \quad \Omega_p \tag{6.73}$$

$$g_j = -\phi_2 \quad \text{on} \quad \Omega_b \tag{6.74}$$

Equation (6.72) states that the local electric field on Ω_b comes from the applied electric field [25,76]. Note that if Ω_b is a cylindrical or a planar surface, Eq. (6.72) reduces to the following [26,29]:

$$\mathbf{n} \cdot \nabla \phi_2 = -E_\infty \quad \text{on} \quad \Omega_b \tag{6.75}$$

6.5.3.2 Flow Field

The assumption is that both Ω_b and Ω_b are no-slip leads to the following boundary conditions for the flow field:

$$\mathbf{u} = U\mathbf{e}_z \quad \text{on} \quad \Omega_p \tag{6.76}$$

$$\mathbf{u} = 0 \quad \text{on} \quad \Omega_b \tag{6.77}$$

where U is the particle velocity in the z direction.

6.5.4 Forces Acting on a Particle

Solving Eqs. (6.39)–(6.43) or Eqs. (6.45)–(6.48) subject to appropriately assumed boundary conditions yields the spatial distributions of the electrical potential, the concentration of ionic species, and the velocity. These results can then be used to evaluate the electrophoretic mobility of a particle, the velocity of a particle per unit strength of the applied electric field, based on a force

balance [75]. Depending upon the nature of a problem, the solution procedure may involve a trial-and-error step. For illustration, let us focus on the
motion of a particle in the z direction, and therefore only the z components
of the relevant forces are considered. In the present case, the forces acting on
a particle include the electrostatic force and the hydrodynamic force. Under
the conditions that the applied electric field is weak relative to that established by Ω_p and/or Ω_b and the equilibrium electric field has no contribution to the movement of the particle, these forces can be evaluated from an
integration of the Maxwell stress tensor and the hydrodynamic stress tensor
over Ω_p [66].

The z component of the electric force acting on the particle, F_E, can be evaluated by

$$F_E = \iint_{\Omega_p} (\boldsymbol{\sigma}^E \cdot \mathbf{n}) \cdot \mathbf{e}_z \, d\Omega_p \tag{6.78}$$

where $\boldsymbol{\sigma}^E = \varepsilon \mathbf{E}_\infty \mathbf{E}_\infty - (1/2)\varepsilon E_\infty^2 \mathbf{I}$ is the Maxwell stress tensor; $\mathbf{E}_\infty = -\nabla\phi = \mathbf{n}(\partial\phi/\partial n) + \mathbf{t}(\partial\phi/\partial t)$; $\mathbf{I}$ is the unit tensor; $\mathbf{t}$ is the unit tangential vector on Ω_p; n and t are the magnitude of $\mathbf{n}$ and that of $\mathbf{t}$, respectively; and $E_\infty^2 = \mathbf{E}_\infty \cdot \mathbf{E}_\infty$.
Substituting $\boldsymbol{\sigma}^E$ into Eq. (6.78), we obtain

$$F_E = \iint_{\Omega_p} \left(\varepsilon \frac{\partial\phi}{\partial n}\frac{\partial\phi}{\partial z} - \frac{1}{2}\varepsilon\left[\left(\frac{\partial\phi}{\partial n}\right)^2 + \left(\frac{\partial\phi}{\partial t}\right)^2 \right] n_z \right) d\Omega_p \tag{6.79}$$

where n_z is the z component of $\mathbf{n}$. Substituting $\phi = \phi_1 + \phi_2$ into this expression
yields

$$\begin{aligned} F_E = &\iint_{\Omega_p} \left(\varepsilon \frac{\partial\phi_1}{\partial n}\frac{\partial\phi_1}{\partial z} - \frac{1}{2}\varepsilon\left[\left(\frac{\partial\phi_1}{\partial n}\right)^2 + \left(\frac{\partial\phi_1}{\partial t}\right)^2 \right] n_z \right) d\Omega_p \\ &+ \iint_{\Omega_p} \left(\varepsilon \frac{\partial\phi_2}{\partial n}\frac{\partial\phi_2}{\partial z} - \frac{1}{2}\varepsilon\left[\left(\frac{\partial\phi_2}{\partial n}\right)^2 + \left(\frac{\partial\phi_2}{\partial t}\right)^2 \right] n_z \right) d\Omega_p \\ &+ \iint_{\Omega_p} \left(\varepsilon\left[\frac{\partial\phi_1}{\partial n}\frac{\partial\phi_2}{\partial z} + \frac{\partial\phi_2}{\partial n}\frac{\partial\phi_1}{\partial z} \right] - \varepsilon\left[\frac{\partial\phi_1}{\partial n}\frac{\partial\phi_2}{\partial n} + \frac{\partial\phi_1}{\partial t}\frac{\partial\phi_2}{\partial t} \right] n_z \right) d\Omega_p \end{aligned} \tag{6.80}$$

Because a particle cannot be driven by its equilibrium electrostatic field, the
terms associated with $(\partial\phi_1/\partial n)^2$, $(\partial\phi_1/\partial t)^2$, and $(\partial\phi_1/\partial n)(\partial\phi_1/\partial z)$ must vanish. In addition, $(\partial\phi_2/\partial n)^2$, $(\partial\phi_2/\partial t)^2$, and $(\partial\phi_2/\partial n)(\partial\phi_2/\partial z)$ are small because

they all involve the product of two perturbed terms. Therefore, Eq. (6.80) reduces to the following [66]:

$$F_E = \iint_{\Omega_p} \left(\varepsilon \left[\frac{\partial \phi_1}{\partial n} \frac{\partial \phi_2}{\partial z} + \frac{\partial \phi_2}{\partial n} \frac{\partial \phi_1}{\partial z} \right] - \varepsilon \left[\frac{\partial \phi_1}{\partial n} \frac{\partial \phi_2}{\partial n} + \frac{\partial \phi_1}{\partial t} \frac{\partial \phi_2}{\partial t} \right] n_z \right) d\Omega_p \qquad (6.81)$$

Note that the magnitude of each term in the integrand of this expression is on the order of E_∞. Equation (6.81) is applicable to both symmetric and asymmetric problems at an arbitrary level of surface potential subject to all types of surface conditions. There are three symmetric axes in a symmetric system. The contours of the electric potential (equal potential surface) in such a system are symmetric, that is, $\partial \phi_1 / \partial t = 0$. Examples of a symmetric system include a sphere at the center of a spherical cavity [25,27,28], and a cylinder [34,44,64], spheroid [36,38,60] or sphere [26,29,40,55] moving along the axis of a cylindrical pore. In contrast, the contours of the electric potential are asymmetric in an asymmetric system. Examples of such a system include a sphere moving normal to a planar surface [14–21,31,32,46], a sphere at an arbitrary position in a spherical cavity [35,37,42], and several other types of geometry [23,39,43,48–51].

For a nonconductive particle, $\partial \phi_2 / \partial n = 0$, and Eq. (6.81) becomes

$$F_E = \iint_{\Omega_p} \left(\varepsilon \left[\frac{\partial \phi_1}{\partial n} \frac{\partial \phi_2}{\partial z} \right] - \varepsilon \left[\frac{\partial \phi_1}{\partial t} \frac{\partial \phi_2}{\partial t} \right] n_z \right) d\Omega_p \qquad (6.82)$$

If Ω_p is maintained at a constant surface potential, $\partial \phi_1 / \partial t = 0$, and this expression can be simplified further as

$$F_E = \iint_{\Omega_p} \varepsilon \frac{\partial \phi_1}{\partial n} \frac{\partial \phi_2}{\partial z} d\Omega_p = \iint_{\Omega_p} \sigma_p E_z d\Omega_p \qquad (6.83)$$

where $\sigma_p = -\varepsilon \mathbf{n} \cdot \nabla \phi_1 = -\varepsilon (\partial \phi_1 / \partial n)$ is the surface charge density, and $E_z = -\partial \phi_2 / \partial z$ is the strength of the local electric field in the z direction. Equation (6.83) was used by Shugai and Carnie [29] in the study of the electrophoresis of a rigid sphere with a thick double layer parallel or normal to a plane and along the axis of a cylindrical pore, and by Hsu et al. in several types of geometries [34,44,47,48,53,54,73,77,78]. These studies all assumed a weak applied electric field and low constant surface potential. It is interesting to note that for a symmetric system, because $\partial \phi_1 / \partial t = 0$ in Eq. (6.82), and therefore Eq. (6.83) is not limited to the case of constant surface potential, but becomes applicable to other types of surface conditions too. For instance, in the electrophoresis of a particle along the axis of a cylindrical pore [34,44,48,77] and at the center of a spherical cavity [25,27,28], using Eq. (6.83) is appropriate.

The expression below is widely used to evaluate the electrostatic force [7,27,38,40,67,68,75]:

$$F_E = \iint\limits_{\Omega_p} \sigma \left(-\frac{\partial \phi}{\partial z} \right) d\Omega_p \tag{6.84}$$

where $\sigma = -\varepsilon \mathbf{n} \cdot \nabla \phi = -\varepsilon (\partial \phi / \partial n)$ is the surface charge density. In terms of ϕ_1 and ϕ_2, this expression becomes

$$F_E = \iint\limits_{\Omega_p} \left(\varepsilon \frac{\partial \phi_1}{\partial n} \frac{\partial \phi_1}{\partial z} + \varepsilon \frac{\partial \phi_1}{\partial n} \frac{\partial \phi_2}{\partial z} + \varepsilon \frac{\partial \phi_2}{\partial n} \frac{\partial \phi_1}{\partial z} + \varepsilon \frac{\partial \phi_2}{\partial n} \frac{\partial \phi_2}{\partial z} \right) d\Omega_p \tag{6.85}$$

The presence of the first, the third, and the fourth terms in the integrand of this expression implies that it is applicable to symmetric electrophoresis problems only, such as a sphere moving at the center of a spherical cavity [27], and a spheroid [38] and a sphere [40] moving along the axis of a cylindrical pore.

The hydrodynamic force acting on the particle in the z direction, F_D, can be evaluated by the following [79]:

$$F_D = \iint\limits_{\Omega_p} (\boldsymbol{\sigma}^H \cdot \mathbf{n}) \cdot \mathbf{e}_z \, d\Omega_p \tag{6.86}$$

where $\boldsymbol{\sigma}^H = -p\mathbf{I} + 2\eta\Delta$ is the hydrodynamic stress tensor; $\Delta = [\nabla\mathbf{u} + (\nabla\mathbf{u})^{\mathsf{T}}]/2$ is the rate of deformation tensor; and the superscript $\mathbf{T}$ denotes matrix transposition. Note that Eq. (6.86) is suitable for both rigid [14–18] and soft particles [41,67]. If we let $\mathbf{u} = \mathbf{n}u_n + \mathbf{t}u_t = \mathbf{n}(\mathbf{u} \cdot \mathbf{n}) + \mathbf{t}(\mathbf{u} \cdot \mathbf{t})$, then Eq. (6.86) can be rewritten as

$$F_D = \iint\limits_{\Omega_p} \left[t_z \eta \left(\frac{\partial u_n}{\partial t} + \frac{\partial u_t}{\partial n} \right) + n_z \left(-p + 2\eta \frac{\partial u_n}{\partial n} \right) \right] d\Omega_p \tag{6.87}$$

For a rigid particle, because $(\partial u_n / \partial n)_{\Omega_p} = 0$ and $(\partial u_n / \partial t)_{\Omega_p} = 0$, this expression can be simplified further as

$$F_D = \iint\limits_{\Omega_p} \eta \frac{\partial (\mathbf{u} \cdot \mathbf{t})}{\partial n} t_z \, d\Omega_p + \iint\limits_{\Omega_p} -p n_z \, d\Omega_p \tag{6.88}$$

This expression indicates that F_D comprises a viscous component and a pressure component [34,80]. The electrophoretic mobility of a particle can be determined from the fact that the net force acting on it in the z direction vanishes at steady state, that is,

$$F_E + F_D = 0 \qquad\qquad (6.89)$$

6.6 Solution Procedure

In this section, approaches for the resolution of the governing equations of electrophoresis, the electrokinetic equations, including both analytical and numerical methods, are introduced for the case where a boundary is present, followed by a brief discussion on the boundary effect on electrophoresis.

6.6.1 Analytical Solution

Due to its nonlinear and coupled nature, solving the set of electrokinetic equations analytically is almost impossible under general conditions. Therefore, simplifications of the original problem so that analytical solutions can be derived are inevitable. However, care must be taken in simplifying the original problem such that the qualitative nature of the phenomenon under consideration is maintained. Otherwise, the results obtained are of limited value. If the double layer surrounding a charged particle takes a finite thickness, that nature includes three key components [81], namely, electrophoretic retardation, the relaxation or polarization effect, and surface conductance. The electrophoretic retardation comes from the relative motion of a charged particle and the associated double layer. A charged particle in an electrolyte solution is surrounded by an ionic cloud (or double layer) richer in counterions. If an external electric field of strength E_∞ is applied, the ionic cloud creates an electric body force of magnitude $-\rho_e E_\infty$ in the fluid, with ρ_e being the effective volumetric charge density. Because the sign of the particle charge is opposite to that of the net charge of the ionic cloud, the direction of the particle motion when the external electric field is applied is also opposite to that of the overall fluid motion. Therefore, the movement of the particle will be retarded by that of the surrounding fluid; this is known as *electrophoretic retardation*. The electrophoretic retardation effect can be taken into account by considering the governing equations, which include the electric body force term, for the flow field.

In the absence of an external electric field, the double layer surrounding a charged spherical particle is symmetric (i.e., spherical). This symmetry may no longer exist as the particle moves subject to an applied external electric field. Because the permittivity and the conductivity of a rigid particle are

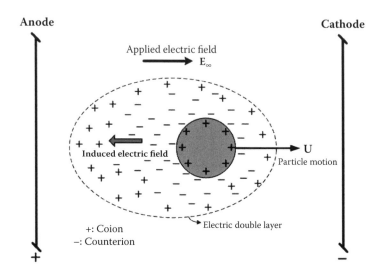

FIGURE 6.5
Distortion of a double layer due to an applied external electric field.

different from those of the surrounding liquid, in general, induced charge and an associated electric field are present on the particle surface when an external electric field is applied. The movement of the particle driven by the applied electric field relative to the mobile ions near the particle surface leads to an asymmetric double layer. A schematic representation of an asymmetric double layer surrounding a positively charged spherical particle is shown in Figure 6.5.

A finite time interval, defined as the relaxation time, is required for the distorted double layer to go back to the original symmetric status through ionic convection, diffusion, and migration. As illustrated in Figure 6.5, the asymmetric double layer yields an internal electric field, and because the direction of it is opposite to that of the external electric field, the mobility of the particle is reduced accordingly; this is known as the *relaxation effect*. As will be introduced in this chapter, the relaxation effect is reflected by non-zero Peclet numbers for ionic species j, $Pe_j = \varepsilon(z_1 e/k_B T)^2/\eta D_j$, in the Nernst-Planck equation describing the conservation of ionic species j. The relaxation effect is important only if the surface (zeta) potential is sufficiently high. The critical levels are circa 125 mV, 62.5 mV, and 42.5 mV for an aqueous KCl, $Ba(NO)_3$, and $La(Cl)_3$ electrolyte solution, respectively [11], and the thickness of the double layer surrounding a particle is on the order of its linear size, that is, $\kappa a \cong 0.1$. The relaxation effect becomes unimportant for either $\kappa a < 0.1$ or $\kappa a \gg 1$ [11,28].

The presence of the double layer surrounding a charged particle yields a higher electric conductivity near the particle surface than that of the bulk liquid phase. If the surface potential of the particle is low and the double layer is thick, the electric conductivity in the double layer is close to that of the bulk electrolyte. On the other hand, if this is not the case, then the electrophoretic mobility of the particle will be affected by the difference in the electric conductivity between the two, known as the *effect of surface conductance*. If this effect is important, the calculated zeta potential can be much lower than that for the case where it is unimportant [81].

As mentioned previously, because solving the general governing equations, Eqs. (6.7)–(6.11), which account for all the above effects, is nontrivial, reported analytical results are based on conditions where one or more of those effects can be neglected. The analysis is usually begun by assuming that the external electric field is relatively weak compared with that established by the charged particle, that is, the applied electric field yields but a small perturbation in the value of each dependent variable compared with the corresponding equilibrium value. Depending upon how the perturbation is applied, the resultant analytical expressions and the corresponding applicable ranges can be different. In the following discussion, the solution procedure introduced above is applied first to the simplest case, that is, the electrophoresis of an isolated spherical particle under the conditions of low surface potential and infinitely thick ($\kappa a \to 0$) [5] or infinitely thin ($\kappa a \to \infty$) [4] double layers. The problem considered by Henry [6], where the thickness of the double layer is arbitrary ($0 < \kappa a < \infty$), is then solved. Note that the analysis of Henry [6] is limited to the case of low surface potential, a negligible relaxation effect arising from ionic convection, and the absence of a boundary effect. A brief overview of the analysis of electrophoresis when the presence of a boundary is significant is also presented in this section using the idealized sphere-in-spherical cavity geometry considered by Zydney [25] as an example.

6.6.1.1 Electrophoresis of an Isolated Sphere in the Limit $\kappa a \ll 1$

Referring to Figure 6.6, let us consider the electrophoresis of a rigid, nonconducting, spherical particle of radius a in an infinite electrolyte solution. A uniform electric field $\mathbf{E}_\infty$ of strength E_∞ in the z direction is applied. Suppose that the concentrations of the ionic species in the liquid phase are sufficiently low so that the Debye length κ^{-1} is very large compared to the particle radius, that is, $\kappa a \ll 1$.

If the spherical coordinates are adopted with its origin placed at the center of the sphere, then the present problem is of one-dimensional nature, where the only independent variable is the radial distance, r. Assuming a symmetric Z:Z electrolyte, the governing Poisson-Boltzmann equation for the electric field, Eq. (6.39), in the Debye-Hückel limit becomes Eq. (6.45), rewritten as below, for convenience:

$$\frac{1}{r^2} \frac{d}{dr}\left(r^2 \frac{d\phi}{dr} \right) = \kappa^2 \phi \tag{6.90}$$

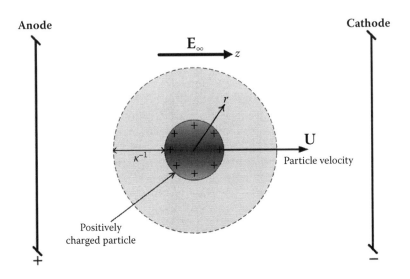

FIGURE 6.6
Electrophoresis of a spherical particle for the case of a thick double layer ($\kappa a \ll 1$).

where $\kappa = (2e^2Z^2n_{10}/\epsilon k_B T)^{1/2}$ is the inverse Debye length. Making the transformation

$$\xi = r\phi \tag{6.91}$$

Eq. (6.90) reduces to

$$\frac{d^2\xi}{dr^2} = \kappa^2\xi \tag{6.92}$$

The general solution to this equation is

$$\xi = C_1 e^{-\kappa r} + C_2 e^{\kappa r} = r\phi \tag{6.93}$$

The constants C_1 and C_2 need to be evaluated from prespecified boundary conditions. Suppose that

$$\phi = \phi_{\Omega_p} \cong \zeta_p \quad \text{at} \quad r = a \tag{6.94}$$

and

$$\phi = \phi_\infty = 0 \quad \text{as} \quad r \to \infty \tag{6.95}$$

Note that the surface potential ϕ_{Ω_p} is approximated by the corresponding zeta potential, ζ_p. After applying Eqs. (6.94) and (6.95), we obtain the following potential distribution:

$$\phi = \zeta_p \frac{a}{r} \exp[-\kappa(r-a)] \tag{6.96}$$

Considering a stationary diffuse double layer surrounding a rigid, non-conducting, spherical charged particle of radius a, the total amount of free charge in the liquid phase, Q_e, can be expressed as

$$Q_e = -\int_a^\infty (4\pi r^2)\rho_e dr \tag{6.97}$$

For the present case, the electrical potential can be described by the Poisson equation

$$\frac{1}{r^2}\frac{d}{dr}\left(r^2\frac{d\phi}{dr}\right) = -\frac{\rho_e}{\varepsilon} \tag{6.98}$$

Substituting this expression into Eq. (6.97) yields

$$Q_e = -4\pi\varepsilon\int_a^\infty r^2\left[\frac{1}{r^2}\frac{d}{dr}\left(r^2\frac{d\phi}{dr}\right)\right]dr = -4\pi\varepsilon\left[r^2\frac{d\phi}{dr}\right]_a^\infty \tag{6.99}$$

Because $(d\phi/dr) \to 0$ as $r \to \infty$, the above expression reduces to

$$Q_e = 4\pi a^2\varepsilon\frac{d\phi}{dr}\bigg|_{r=a} = -4\pi a\varepsilon(1+\kappa a)\zeta_p \tag{6.100}$$

Because the condition of electroneutrality must be satisfied, that is, the total amount of charge on the particle surface must be balanced by that in the liquid phase, we have

$$Q_p = -Q_e = 4\pi a\varepsilon(1+\kappa a)\zeta_p \tag{6.101}$$

where Q_p is the total amount of the surface charge of the particle. The surface charge density of the particle, σ_p, can be expressed as

$$\sigma_p = \frac{Q_p}{4\pi a^2} = \varepsilon\left(\frac{1+\kappa a}{a}\right)\zeta_p \tag{6.102}$$

That is, under the Debye-Hückel condition, the surface charge density of a rigid particle linearly correlates with its surface potential.

For the case where the double layer is thick, $\kappa a \ll 1$, Eq. (6.101) becomes

$$Q_p = 4\pi a \varepsilon \zeta_p \tag{6.103}$$

Therefore, the electrical force acting on the particle when an external field E_∞ is applied is

$$F_E = Q_p E_\infty = 4\pi a \varepsilon \zeta_p E_\infty \tag{6.104}$$

At steady state, this force is balanced by the hydrodynamic force acting on the particle, $F_D = 6\pi\eta U a$ [79], where U is the electrophoretic velocity of the particle, yielding

$$Q_p E_\infty = 6\pi\eta U a \tag{6.105}$$

The electrophoretic velocity of the particle can be determined by this expression and Eq. (6.104) as

$$U = \frac{2}{3}\frac{\varepsilon \zeta_p E_\infty}{\eta} \tag{6.106}$$

Recall that this expression is valid for $\kappa a \ll 1$. Equation (6.106) is usually expressed as Eq. (6.2), $\mu = U/E_\infty = (2/3)(\varepsilon\zeta_p/\eta)$, which is Hückel's formula for the electrophoretic mobility of a particle.

6.6.1.2 Electrophoresis of an Isolated Sphere in the Limit $\kappa a \gg 1$

If the particle is large or the double layer surrounding it is thin so that $\kappa a \gg 1$, the curvature of the particle surface can be neglected, that is, it can be treated as a planar surface. In this case, the electrophoresis problem can be treated as the flow of an electrolyte solution past a planar surface with the direction of the applied electric field parallel to the surface, as illustrated in Figure 6.7.

As shown in Figure 6.7, if we choose the Cartesian coordinates with the origin on the particle surface, then the fluid velocity on that surface vanishes and the fluid velocity far away from that surface is $-U$. In this case, Eq. (6.9) becomes

$$\eta \frac{d^2 u_z}{dy^2} = -\rho_e E_\infty \tag{6.107}$$

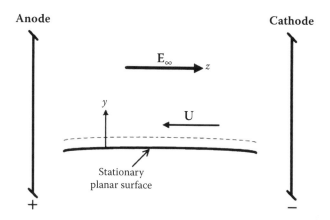

FIGURE 6.7
Electrophoresis of a particle for the case of a thin double layer ($\kappa a \gg 1$).

where u_z is the velocity of the fluid tangential to the particle surface. Combining with the Poisson equation,

$$\frac{d^2\phi}{dy^2} = -\frac{\rho_e}{\varepsilon}$$

(6.108)

Eq. (6.107) becomes

$$\eta \frac{d^2 u_z}{dy^2} = \varepsilon E_\infty \frac{d^2\phi}{dy^2}$$

(6.109)

The equation needs to be solved subject to the following boundary conditions:

$$\phi \to 0 \quad \text{and} \quad u_z \to -U \quad \text{as} \quad y \to \infty$$

(6.110)

$$\frac{d\phi}{dy} \to 0 \quad \text{and} \quad \frac{du_z}{dy} \to 0 \quad \text{as} \quad y \to \infty$$

(6.111)

$$\phi = \zeta_p \quad \text{and} \quad u_z = 0 \quad \text{at} \quad y = 0$$

(6.112)

Integrating Eq. (6.109) from infinity to an arbitrary distance y from the particle surface and employing the boundary conditions, Eqs. (6.110) and (6.111), we obtain

$$u_z(y) = \frac{\varepsilon E_\infty \phi(y)}{\eta} - U$$

(6.113)

Substituting the boundary condition Eq. (6.112) into this expression yields

$$U = \frac{\varepsilon E_\infty \zeta_p}{\eta} \tag{6.114}$$

This expression can be rewritten as $\mu = U/E_\infty = \varepsilon \zeta_p/\eta$, which is Helmholtz-Smoluchowski's formula for the electrophoretic mobility of a particle, which is valid for $\kappa a \gg 1$ and $Ze\zeta_p/k_B T < 1$.

Again, the results of Hückel [5] and Helmholtz-Smoluchowski's formula [4], Eqs. (6.106) and (6.114), respectively, represent the limiting cases of infinitely thick ($\kappa a \to 0$) and infinitely thin ($\kappa a \to \infty$) double layers surrounding an isolated particle, respectively. They differ by a factor of 2/3, but both are independent of particle size. Furthermore, because the result of Helmholtz and Smoluchowski is based on a planar surface, that is, the curvature of the particle surface is neglected, it is independent of the shape of the particle, as long as the double layer is sufficiently thin.

6.6.1.3 Electrophoresis of an Isolated Sphere for $0 < \kappa a < \infty$

The results of Hückel and of Helmholtz-Smoluchowski's formula were bridged by Henry's formula [6] in 1931 for intermediate values of κa. That formula can be derived from the linear equations for the perturbation dependent variable, Eqs. (6.45)–(6.48), under the condition of low surface potential.

Let us consider the electrophoresis of a rigid, positively charged, nonconducting spherical particle of radius a in an infinite electrolyte solution subject to an applied uniform electric field $\mathbf{E}_\infty$ of strength E_∞ in the z direction. In this case, the electrophoretic velocity of the particle U is in the positive z direction. For convenience, the particle is held fixed and the liquid phase flows with a relative bulk velocity U in the negative z direction, as illustrated in Figure 6.8.

The present problem is one-dimensional. If the spherical coordinates are chosen, then the equilibrium potential described by Eq. (6.45) becomes

$$\frac{1}{r^2}\frac{d}{dr}\left(r^2 \frac{d\phi_1}{dr} \right) = \kappa^2 \phi_1 \tag{6.115}$$

The associated boundary conditions are

$$\phi_1 = \zeta_p \quad \text{at} \quad r = a \tag{6.116}$$

$$\phi_1 \to 0 \quad \text{as} \quad r \to \infty \tag{6.117}$$

It can be shown without too much difficulty that the solution to Eq. (6.115) subject to Eqs. (6.116) and (6.117) is

$$\phi_1 = \zeta_p \frac{a}{r}\exp[-\kappa(r-a)] \tag{6.118}$$

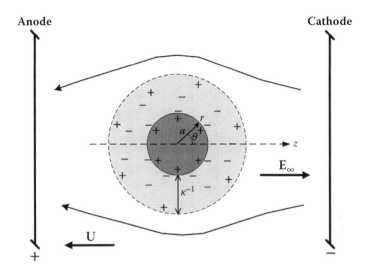

FIGURE 6.8
Electrophoresis of a rigid, positively charged, nonconducting spherical particle of radius a in an infinite electrolyte solution.

Note: $\mathbf{E}_\infty$ is an applied uniform electric field, which is $\mathbf{E}_\infty$ in the z direction. The particle is held fixed, and the liquid phase flows with a relative bulk velocity U in the negative z direction.

Note that this potential comes solely from the charged particle.

The perturbed potential ϕ_2, which arises from the applied electric field, must satisfy the Laplace equation, Eq. (6.46). For the present case, this equation becomes

$$\frac{1}{r^2}\frac{\partial}{\partial r}\left(r^2\frac{\partial\phi_2}{\partial r}\right) + \frac{1}{r^2\sin\theta}\frac{\partial}{\partial\theta}\left(\sin\theta\frac{\partial\phi_2}{\partial\theta}\right) = 0 \qquad (6.119)$$

The assumed boundary conditions associated with Eq. (6.119) are

$$\frac{\partial\phi_2}{\partial r} = 0 \quad \text{at} \quad r = a \qquad (6.120)$$

$$\phi_2 = -E_\infty r\cos\theta \quad \text{as} \quad r \to \infty \qquad (6.121)$$

The boundary condition assumed on the particle surface, Eq. (6.120), is based on the assumption that the effect of surface conductance is negligible. That is, the permittivity of the particle is insignificant compared to that of the surrounding fluid. The boundary condition expressed in Eq. (6.121) implies that the electric potential far away from the particle is not influenced by its presence. This equation suggests that the solution to Eq. (6.119) takes the form

$$\phi_2 = f(r)\cos\theta \tag{6.122}$$

Substituting this expression into Eq. (6.119) yields

$$2r\frac{df}{dr} + r^2\frac{df}{dr} - 2f = 0 \tag{6.123}$$

The general solution to this equation is

$$f = \left(\frac{C_3}{r^2} + C_4 r\right) \tag{6.124}$$

where the arbitrary constants C_3 and C_4 need to be determined from the boundary conditions Eqs. (6.120) and (6.121). It can be shown that the final result is

$$f = \left(\frac{-a^3 E_\infty}{2}\right)\frac{1}{r^2} + (-E_\infty)r \tag{6.125}$$

and

$$\phi_2 = -E_\infty\left(r + \frac{a^3}{2r^2}\right)\cos\theta \tag{6.126}$$

In our problem, the forces acting on the particle include the electric force and the hydrodynamic force. The geometric nature of the problem indicates that knowing the z components of these forces is sufficient. To evaluate the hydrodynamic force, the flow field around the sphere has to be solved. For creeping flow at steady state, this field is described by the continuity equation and the modified Navier-Stokes equation:

$$\nabla \cdot \mathbf{u} = 0 \tag{6.127}$$

$$\eta\nabla^2\mathbf{u} - \nabla p = -\varepsilon\nabla^2\phi_1\nabla(\phi_1 + \phi_2) \tag{6.128}$$

where ϕ_1 and ϕ_2 are specified by Eqs. (6.118) and (6.126), respectively. Because we assume for convenience that the particle is held fixed and the fluid moves with a bulk relative velocity $-U$, the boundary conditions associated with Eqs. (6.127) and (6.128) are

$$u_r = 0 \quad \text{and} \quad u_\theta = 0 \quad \text{at} \quad r = a \tag{6.129}$$

$$u_r = -U\cos\theta \quad \text{and} \quad u_\theta = U\sin\theta \quad \text{as} \quad r \to \infty \tag{6.130}$$

In these expressions, u_r and u_θ are the r and the θ components of the fluid velocity $\mathbf{u}$. It can be shown that the general solution to Eqs. (6.127) and (6.128) subject to Eqs. (6.129) and (6.130) is as follows [6,79]:

$$u_r = \left(B_1 r^2 + \frac{B_2}{r} + B_3 + \frac{B_4}{r^3} \right) \cos\theta + \frac{2\varepsilon E_\infty}{3\eta} \left(\int_a^r \varphi\, dr - \frac{1}{r^3} \int_a^r r^3 \varphi\, dr \right) \cos\theta \quad (6.131)$$

$$u_\theta = \left(-2B_1 r^2 - \frac{B_2}{2r} - B_3 + \frac{B_4}{2r^3} \right) \sin\theta - \frac{2\varepsilon E_\infty}{3\eta} \left(\int_a^r \varphi\, dr + \frac{1}{2r^3} \int_a^r r^3 \varphi\, dr \right) \sin\theta \quad (6.132)$$

$$p = p_0 + \eta \left(10 B_1 r + \frac{B_2}{r^2} \right) \cos\theta - \varepsilon E_\infty \left(3 \frac{\partial \phi_1}{\partial r} - 2\varphi \right) \cos\theta$$
$$- \int_\infty^r \left(\frac{1}{r^2} \frac{\partial}{\partial r} \left(r^2 \frac{\partial \phi_1}{\partial r} \right) \right) \frac{\partial \phi_1}{\partial r} dr \quad (6.133)$$

In these expressions, p, u_r, and u_θ are all functions of r and θ, and

$$\varphi = \frac{\partial \phi_1}{\partial r} + \frac{a^3 r}{2} \int_\infty^r \frac{1}{r^6} \frac{\partial}{\partial r} \left(r^2 \frac{\partial \phi_1}{\partial r} \right) dr \quad (6.134)$$

The integration constants B_i, $i = 1,2,3,4$, need to be determined from the hydrodynamic boundary conditions, Eqs. (6.129) and (6.130).

The hydrodynamic force F_D exerted by the fluid on the particle can be calculated by

$$F_D = 2\pi a^2 \int_0^\pi \left[-\tau_{rr} \cos\theta + \tau_{r\theta} \sin\theta \right]_{r=a} \sin\theta\, d\theta \quad (6.135)$$

Here,

$$\tau_{rr} = -p + 2\eta \frac{\partial u_r}{\partial r} \quad (6.136)$$

and

$$\tau_{r\theta} = \eta \left(\frac{\partial u_\theta}{\partial r} - \frac{u_\theta}{r} + \frac{1}{r} \frac{\partial u_r}{\partial \theta} \right) \quad (6.137)$$

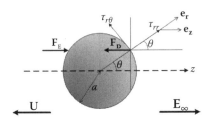

FIGURE 6.9
Hydrodynamic and electric forces acting on a particle surface.

represent the normal and the tangential stresses exerted by the fluid on the particle surface, respectively. Figure 6.9 illustrates the relevant forces on the particle surface.

As stated in the previous section, the electric force acting on the particle in the z direction, F_E, can be evaluated by Eq. (6.83), which is rewritten below for convenience:

$$F_E = \iint_{\Omega_p} [\sigma_p E_z]_{\Omega_p} \, d\Omega_p \qquad (6.138)$$

In this expression, σ_p can be evaluated by

$$\sigma_p = \frac{Q_p}{4\pi a^2} = -\varepsilon \frac{d\phi_1}{dr}\bigg|_{r=a} \qquad (6.139)$$

The strength of the local electric field in the z direction can be expressed by

$$E_z = -E_\theta \sin\theta + E_r \cos\theta \qquad (6.140)$$

where E_r and E_θ are the r and the θ components of E_z, respectively. These components can be evaluated by

$$E_r = -\frac{\partial}{\partial r}[\phi_1(r) + \phi_2(r,\theta)] = -\frac{d\phi_1}{dr} \qquad (6.141)$$

and

$$E_\theta = -\frac{1}{r}\frac{\partial}{\partial\theta}[\phi_1(r) + \phi_2(r,\theta)] = -\frac{1}{r}\frac{\partial\phi_2}{\partial\theta} \qquad (6.142)$$

Note that Eq. (6.141) comes from the boundary equation specified on the particle surface, Eq. (6.120). Substituting Eqs. (6.141) and (6.142) into Eq. (6.140) leads to

$$E_z = \frac{1}{r}\frac{\partial\phi_2}{\partial\theta}\sin\theta - \frac{d\phi_1}{dr}\cos\theta \tag{6.143}$$

Substituting Eqs. (6.139) and (6.143) into Eq. (6.138) yields

$$F_E = \int_0^\pi \left[-\varepsilon\frac{d\phi_1}{dr}\right]\left[\frac{1}{a}\frac{\partial\phi_2}{\partial\theta}\sin\theta - \frac{d\phi_1}{dr}\cos\theta\right][2\pi a\sin\theta]ad\theta \tag{6.144}$$

Because $\int_0^\pi \cos\theta\sin\theta d\theta = 0$, the term $(d\phi_1/dr)\cos\theta$ on the right-hand side of Eq. (6.144) does not contribute to the electric force, as stated in the previous section. That is, the particle cannot be driven by the electric field established by itself, the equilibrium electric field corresponding to the equilibrium potential. Therefore, Eq. (6.144) reduces to

$$F_E = -2\pi a\varepsilon\frac{d\phi_1}{dr}\bigg|_{r=a}\int_0^\pi \frac{\partial\phi_2}{\partial\theta}\bigg|_{r=a}\sin^2\theta d\theta \tag{6.145}$$

Differentiating Eq. (6.126) with respect to θ and letting $r = a$ in the resultant expression, we obtain

$$\frac{d\phi_2}{d\theta}\bigg|_{r=a} = \frac{3a}{2}E_\infty\sin\theta \tag{6.146}$$

Substituting this expression into Eq. (6.145), we obtain

$$F_E = \left[-4\pi a^2\varepsilon\frac{d\phi_1}{dr}\bigg|_{r=a}\right]E_\infty \tag{6.147}$$

which is the electric force acting on a particle in the z direction in the electrophoresis problem considered by Henry [6]. The term in the square brackets on the right-hand side of Eq. (6.147) represents the total surface charge. Note that in the present case, the perturbed potential ϕ_2 does not contribute to the electric force F_E because it acts on a charge-free dielectric. However, this may not be the case if the presence of a boundary (boundary effect) and/or a neighboring particle (concentration effect) is important.

At steady state, the electric force acting on the particle is balanced by the corresponding hydrodynamic force, that is,

$$F_E = F_D \tag{6.148}$$

which leads to

$$U = \left[\frac{2\varepsilon\zeta_p E_\infty}{3\eta} \right] f(\kappa a) \tag{6.149}$$

where $f(\kappa a)$ is Henry's function [6]. Equation (6.149) is usually rearranged as $\mu = U/E_\infty = (2\varepsilon\zeta_p/3\eta) f(\kappa a)$, which is Henry's formula.

Henry's formula provides a very convenient and useful analytical expression for the electrophoretic mobility of an isolated spherical particle valid for an arbitrary double-layer thickness, provided that the surface potential is low $(Ze\zeta_p/k_B T < 1)$. However, because the boundary effect and the concentration effect are neglected, it becomes inaccurate if the electrophoresis is conducted, for instance, in a narrow space and/or the concentration of particles is appreciable. To take the boundary effect into account, several analyses are begun by assuming thin double layers, which is shown to be appropriate if the following applies [11,82]:

$$(\kappa a)^{-1} \cosh\left(\frac{Ze\zeta_p}{2k_B T} \right) << 1 \tag{6.150}$$

For instance, in a study of various types of boundary effect, Keh and Anderson [13] assumed that the double layer surrounding a spherical particle and/or the boundary is infinitely thin (i.e., $\kappa a \rightarrow \infty$). The governing equations for the electric, the flow, and the concentration fields can be simplified and reorganized through scaling arguments. The domain under consideration is partitioned into an inner region and an outer region. The former ranges from the particle surface to a liquid surface, which is κ^{-1} from the particle surface. The latter ranges from the outer boundary of the inner region to the system boundary. The inner region is essentially the double layer. Under the condition of low surface potential, that is, ζ_p is on the order of $k_B T/e$, the governing electrokinetic equations for the inner region are

$$\nabla^2 \phi_{(i)} = \kappa^2 \phi_{(i)} \tag{6.151}$$

$$\nabla \cdot \mathbf{u}_{(i)} = 0 \tag{6.152}$$

$$\eta\nabla^2 \mathbf{u}_{(i)} - \rho_e \nabla\phi_{(i)} = 0 \tag{6.153}$$

In these expressions, the subscript (i) denotes the properties in the inner region, and $\rho_e = -\varepsilon\nabla^2\phi_{(i)}$. Note that Eqs. (6.151)–(6.153) are linear and decoupled. In addition, because the double layer is very thin compared with the particle radius, the curvature of the particle surface becomes unimportant, and therefore, the solid–fluid interface is assumed planar. In this case,

Eqs. (6.151)–(6.153) can be integrated directly to provide the boundary conditions for the outer region. This approach is the same as that adopted by O'Brien [83].

In the outer region, applying the condition of electroneutrality, that is, ρ_e vanishes in that region, leads to the following governing equations:

$$\nabla^2 \phi_{(o)} = 0 \tag{6.154}$$

$$\nabla \cdot \mathbf{u}_{(o)} = 0 \tag{6.155}$$

$$\eta \nabla^2 \mathbf{u}_{(o)} - \nabla p_{(o)} = 0 \tag{6.156}$$

Here, the subscript (o) denotes the properties in the outer region. As in the case of the inner region, the governing electrokinetic equations in the outer region are also linear and decoupled. Keh and Anderson [14] were able to derive the expression below using the method of reflections (asymptotic solution) for the electrophoresis of a charged nonconducting sphere parallel to a nonconductive plate (case 1), perpendicular to a dielectric plane (case 2), along the centerline between two parallel plates (case 3), and along the axis of a cylindrical pore (case 4):

$$U = f_i(\lambda)(\zeta_p - \zeta_b)\frac{\varepsilon E_\infty}{\eta}, \quad i = 1, 2, 3, 4 \tag{6.157}$$

where $f_i(\lambda)$ is a function reflecting the geometry considered. For cases 1 through 4, it can be expressed as

$$f_1(\lambda) = \left[1 - \frac{1}{16}\lambda^3 + \frac{1}{8}\lambda^5 - \frac{3}{256}\lambda^6 + O(\lambda^8)\right], \tag{6.158}$$

$$f_2(\lambda) = \left[1 - \frac{5}{8}\lambda^3 + \frac{1}{4}\lambda^5 - \frac{5}{8}\lambda^6 + O(\lambda^8)\right], \tag{6.159}$$

$$f_3(\lambda) = [1 - 0.267699\lambda^3 + 0.338324\lambda^5 - 0.040224\lambda^6 + O(\lambda^8)], \tag{6.160}$$

$$f_4(\lambda) = [1 - 1.28987\lambda^3 + 1.89632\lambda^5 - 1.02780\lambda^6 + O(\lambda^8)] \tag{6.161}$$

In these expressions, $\lambda = a/b$, with a and b being the radius of the sphere and the distance from the center of the sphere to the boundary, respectively. The influence of a boundary on the electrophoretic behavior of a sphere can be summarized as follows [14]: (a) except for the case of a sphere normal to a conducting plane, the behavior of the sphere is influenced by the electroosmotic flow induced by a charged boundary, and the electrophoretic velocity of the sphere is proportional to the difference in the zeta potentials of the sphere and the boundary; (b) the presence of the boundary has the effect of altering the electrical interaction between the sphere and the applied electric field; and (c) due to the nonslip condition on the boundary, the viscous retardation on a sphere is enhanced. Many of the subsequent analyses were based on the assumption of an infinitely thin double layer, where the influence of the electric field on the flow field is taken into account by incorporating a slip boundary condition into the equations governing the flow field. For instance, assuming infinitely small Debye lengths ($\kappa^{-1} \to 0$), Keh and Chen [15] derived an analytical solution for the electrophoresis of a sphere parallel to a planar wall using spherical bipolar coordinates, for all values of λ (=a/b, with a and b being the particle radius and the distance between the center of the sphere and the plane, respectively). For $\lambda < 0.7$, the results obtained are in good agreement with those of Keh and Anderson [14], which were based on the method of reflections. Due to the deformation of the electric field in the gap between the sphere and the plane, the electrophoretic velocity at very large λ in the present system is somewhat greater than that in an infinite medium. The analysis of Keh and Chen [15] was extended by Keh and Lien [16] to that for the electrophoresis of a sphere along the axis of a circular disk, also limited to an infinitely thin double layer.

6.6.1.4 Electrophoresis of a Sphere in a Spherical Cavity

In this section, the problem considered by Zydney [25], the electrophoresis of a sphere at the center of a spherical cavity, is adopted to illustrate the analytical solution procedure. This study extends Henry's analysis for the electrophoresis of an isolated sphere in an infinite medium to the case where the boundary can be significant. The geometry considered is simple enough so that an analytical expression for the electrophoretic mobility of a particle can be derived. Although this geometry, which is of a one-dimensional nature, is an idealized one, it is capable of simulating the key factors of the electrophoresis of a particle when the boundary effect is significant.

Referring to Figure 6.10, let us consider the electrophoresis of a rigid, nonconducting, spherical particle of radius a at the center of a spherical cavity of radius b. Let $\lambda = a/b$, which measures the significance of the boundary effect. The closer the value of λ to unity, the more significant that effect; and as $\lambda \to 0$, the problem reduces to the electrophoresis of an isolated sphere in an infinite medium. The medium between the sphere and the cavity is filled with an incompressible Newtonian aqueous solution containing 1:1 electrolyte. A uniform electric field $\mathbf{E}_\infty$ of strength E_∞ in the z direction is applied.

Both the surface of the particle and that of the cavity are no-slip, and the electrical potentials of these surfaces are maintained at ζ_p and ζ_b, respectively. Following Henry's treatment [6], the total electrostatic potential ϕ is expressed as the linear superposition of the electrical potential in the absence of $\mathbf{E}_\infty$, ϕ_1, and the potential outside the particle arising from the application

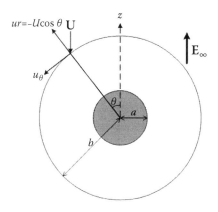

FIGURE 6.10
Electrophoresis of a sphere at the center of a spherical cavity.

of $\mathbf{E}_\infty$, ϕ_2. This approach is valid for the case where $\mathbf{E}_\infty$ is relatively weak compared with that established by the particle and/or the cavity. In addition, both ζ_p and ζ_b are sufficiently low (the Debye-Hückel approximation). Under these conditions, the electrokinetic equations can be expressed as

$$\nabla^2 \phi_1 = \kappa^2 \phi_1 \tag{6.162}$$

$$\nabla^2 \phi_2 = 0 \tag{6.163}$$

$$\nabla \cdot \mathbf{u} = 0 \tag{6.164}$$

$$\eta \nabla^2 \mathbf{u} - \nabla p - \rho_e \nabla(\phi_1 + \phi_2) = 0 \tag{6.165}$$

Here, the spatial density of the mobile ions can be expressed as

$$\rho_e = -\varepsilon \nabla^2 \phi = -\varepsilon \nabla^2(\phi_1 + \phi_2) = -\varepsilon \nabla^2 \phi_1 \tag{6.166}$$

The boundary conditions associated with Eqs. (6.162)–(6.165) are assumed as

$$\phi_1 = \zeta_p \quad \text{at} \quad r = a \tag{6.167}$$

$$\phi_1 = \zeta_b \quad \text{at} \quad r = b \tag{6.168}$$

$$\frac{\partial \phi_2}{\partial r} = 0 \quad \text{at} \quad r = a \tag{6.169}$$

$$\frac{\partial \phi_2}{\partial r} = -E_\infty \cos \theta \quad \text{at} \quad r = a \tag{6.170}$$

$$u_r = 0, \quad \text{and} \quad u_\theta = 0 \quad \text{at} \quad r = a \tag{6.171}$$

$$u_r = -U\cos\theta \quad \text{and} \quad u_\theta = U\sin\theta \quad \text{at} \quad r = b \tag{6.172}$$

Making the transformation $\xi = r\phi_1$, it can be shown that the general solution to Eq. (6.162) can be expressed as

$$\xi = r\phi_1 = C_5\cos h(\kappa r) + C_6\sin h(\kappa r) \tag{6.173}$$

where the arbitrary constants C_5 and C_6 need to be determined from the appropriated boundary conditions expressed in Eqs. (6.167) and (6.168). The final result is

$$\phi_1 = \frac{a\zeta_p}{r}\cos h[\kappa(r-a)] + \frac{b\{\zeta_b - \lambda\zeta_p\cos h[\kappa(b-a)]\}}{r\sin h[\kappa(b-a)]}\sin h[\kappa(b-a)] \tag{6.174}$$

Following a similar procedure as that employed for the resolution of ϕ_1, it can be shown that

$$\phi_2 = -\frac{E_\infty}{1-\lambda^3}\left(r + \frac{a^3}{2r^2}\right)\cos\theta \tag{6.175}$$

Using these electrical potentials, the governing equations for the flow field can be solved through adopting Henry's approach [6] to obtain

$$u_r = \left(A_1 r^2 + \frac{A_2}{r} + A_3 + \frac{A_4}{r^3}\right)\cos\theta + \frac{2X}{3\eta}\left(\int_a^r \psi\,dr - \frac{1}{r^3}\int_a^r r^3\psi\,dr\right)\cos\theta \tag{6.176}$$

$$u_\theta = \left(-2A_1 r^2 - \frac{A_2}{2r} - A_3 + \frac{A_4}{2r^3}\right)\sin\theta - \frac{2X}{3\eta}\left(\int_a^r \psi\,dr + \frac{1}{2r^3}\int_a^r r^3\psi\,dr\right)\sin\theta \tag{6.177}$$

$$p = p_0 + \eta\left(10A_1 r + \frac{A_2}{r^2}\right)\cos\theta - X\left(3\frac{\partial\phi_1}{\partial r} - 2\psi\right)\cos\theta$$

$$-\int_b^r\left(\frac{1}{r^2}\frac{\partial}{\partial r}\left(r^2\frac{\partial\phi_1}{\partial r}\right)\right)\frac{\partial\phi_1}{\partial r}dr \tag{6.178}$$

In these expressions,

$$X = \frac{\varepsilon E_\infty}{1-\lambda^3} \tag{6.179}$$

$$\psi = \frac{\partial\phi_1}{\partial r} + \frac{a^3 r}{2}\int_b^r \frac{1}{r^6}\frac{\partial}{\partial r}\left(r^2\frac{\partial\phi_1}{\partial r}\right)dr \tag{6.180}$$

The integration constants A_i, $i = 1,2,3,4$, need to be determined from the hydrodynamic boundary conditions, Eqs. (171) and (172).

For the present problem, the forces acting on the particle include the electric force and the hydrodynamic force. Due to the symmetric nature of the geometry, only the z components of these forces need to be evaluated. Let F_z, F_E, and F_D be the z components of the total force, the electrostatic force, and the hydrodynamic force acting on the particle, respectively. Then $F_z = F_E + F_D$, with

$$F_E = \sigma_p X = -4\pi a^2 X \left(\frac{d\phi_1}{dr} \right)_{r=a} \tag{6.181}$$

$$F_D = 2\pi a^2 \left[\int_0^\pi \tau_{r\theta}|_{r=a} \sin^2 \theta d\theta + \int_0^\pi \left(-p + 2\eta \frac{\partial u_r}{\partial r} \right)_{r=a} \sin \theta \cos \theta d\theta \right] \tag{6.182}$$

In these expressions, $\tau_{r\theta}|_{r=a}$ is the shear stress evaluated at the particle surface. Substituting the velocity and the pressure profiles, Eqs. (6.176)–(6.178) into Eqs. (6.181) and (6.182), we obtain

$$F_z = -4\pi\eta A_2$$
$$= -6\pi\eta a \frac{(1-\lambda^5)(U+\alpha) - (5/2)(1-\lambda^2)\beta}{1 - (9/4)\lambda + (5/2)\lambda^3 - (9/4)\lambda^5 + \lambda^6} \tag{6.183}$$

with

$$\alpha = \frac{2X}{3\eta} \int_a^b \psi \, dr$$
$$= \frac{2X}{3\eta} \left[-\frac{3}{2}\zeta_p + \zeta_b \left(1 - \lambda^3 + \frac{3}{2}\lambda^5 \right) - \int_\lambda^1 \left(\frac{3\lambda^3}{\xi^4} - \frac{15\lambda^5}{2\xi^6} \right) \phi_1 \, d\xi \right] \tag{6.184}$$

$$\beta = \frac{2X}{3\eta} \int_a^b \frac{r^3}{b^3} \psi \, dr$$
$$= \frac{2X}{3\eta} \left[-\frac{3}{2}\lambda^3 \zeta_p + \zeta_b \left(1 - \frac{\lambda^3}{10} + \frac{3}{5}\lambda^8 \right) - 3\int_\lambda^1 \left(\xi^2 - \frac{\lambda^8}{\xi^6} \right) \phi_1 d\xi \right] \tag{6.185}$$

The electrophoretic velocity of the particle, U, can then be evaluated by letting $F_z = 0$ in Eq. (6.183), and solving the resultant expression to obtain

$$U = \frac{-\alpha(1-\lambda^5) + (5/2)\beta(1-\lambda^2)}{1-\lambda^5} \tag{6.186}$$

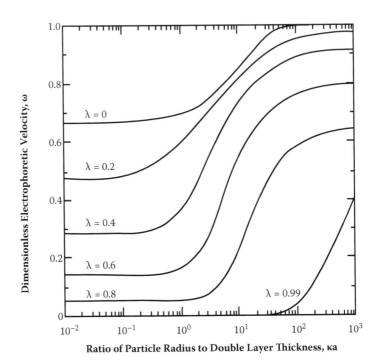

FIGURE 6.11

Variation of the dimensionless electrophoretic mobility ω as a function of κa at various values of λ for the case of a positively charged sphere in an uncharged spherical cavity.

Source: Zydney, A. L., Boundary effects on the electrophoretic motion of a charged particle in a spherical cavity, *J. Colloid Interface Sci.*, 169, 476–485, 1995.

This expression is valid for an arbitrary thickness of double-layer and arbitrary particle sizes. Note, however, that because further integrations are still needed for α and β, Eq. (6.186) is not an explicit analytical expression.

Some features of the present boundary effect deserve a close look. Figure 6.11 shows the variation of the dimensionless electrophoretic mobility, $\omega = \eta U/\varepsilon\zeta_p E_\infty$, with η and E_∞ being the fluid viscosity and the strength of the applied electric field, respectively, as a function of the thickness of the double layer, measured by κa, at various scaled particle radius λ for the case of a positively charged sphere at the center of an uncharged spherical cavity. This figure reveals that the results for $\lambda = 0$ are identical to those presented by Henry [6] for the electrophoresis of an isolated sphere in an infinite solution. In addition, ω varies from (2/3) at small κa (the Hückel limit) to unity at large κa (the Smoluchowski limit). The dependence of the mobility of the particle on the thickness of the double layer is more pronounced as the presence of the boundary becomes more important.

The influence of the cavity on the electrophoretic behavior of the particle is illustrated in Figure 6.12, where the dimensionless mobility ω is plotted

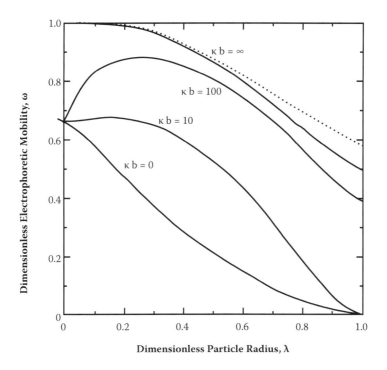

FIGURE 6.12

Variation of the dimensionless electrophoretic mobility ω as a function of the dimensionless particle radius λ at various values of κb.

Note: Dashed line: predicted mobility for a spherical particle in a cylindrical pore at $\kappa b \to \infty$ evaluated by Keh and Anderson [14].

Sources: Keh, H. J., Anderson, J. L., Boundary effects on electrophoretic motion of colloidal spheres, *J. Fluid Mech.*, 153, 417–439, 1985; and Zydney, A. L., Boundary effects on the electrophoretic motion of a charged particle in a spherical cavity, *J. Colloid Interface Sci.*, 169, 476–485, 1995.

against the dimensionless particle radius λ for several values of κa. It was concluded that for $\lambda < 0.7$, the present results are consistent with those of Keh and Anderson [14] for the case of a spherical particle in a cylindrical pore. This consistency, however, is valid for $\lambda < 0.7$, because the results in [14] are limited to the first two reflections. As expected, the boundary effect for the case of thick double layers was found to be much more significant than that for the case of thin double layers. Furthermore, if the cavity is charged, the direction of electrophoresis can be reversed due to the presence of the induced charge on the particle surface and an electroosmotic recirculation flow.

6.6.2 Numerical Solution

A numerical method often becomes the last and necessary choice for the resolution of an electrophoresis problem that is of practical significance. As discussed previously, this occurs, for example, when the double layer surrounding takes a finite thickness, the boundary and/or the concentration effects are significant, or the effect of double-layer polarization needs to be considered. The extension of the classic analysis of Smoluchowski [4] to arrive at an analytical result [4–6,13–18,22,25] is usually based on the conditions of low surface potentials and idealized geometries using linear perturbation approximations. The approximate results thus obtained therefore have many limitations in their applications. For instance, the assumption of low surface potential is adequate only if $|\zeta_p| < 25.4$ mV, which can easily be violated in practice. In that case, instead of solving a linearized Poisson-Boltzmann equation, a nonlinear Poisson-Boltzmann equation needs to be solved. Up to now, the only exactly solvable case is an infinite planar surface of constant electrical potential immersed in a symmetric electrolyte solution. The assumption of an infinitely thin double layer, which is often assumed [13–18,22], is also an idealized condition, which is satisfactory only if a particle is sufficiently large and/or the concentration of ionic species is sufficiently high.

In general, the analysis of an electrophoresis problem involves solving at least three sets of equations, including the equations describing the electric field (the Poisson equation), the flow field (Navier-Stokes and continuity equations), and the ionic concentration field (Nernst-Planck equations). For a three-dimensional system containing N ionic species, there are (N+5) equations: one Poisson equation, three Navier-Stokes equations, one continuity equation, and N Nernst-Planck equations. These coupled, nonlinear, partial differential equations must be solved simultaneously. The Poisson equation involves the space distributions of the electric potential and the ionic species. The latter needs to be determined from the Nernst-Planck equations, which in turn require the solutions of the Navier-Stokes equations, the continuity equation, and the Poisson equation. The Navier-Stokes equations contain the electrical body force, which needs to be determined from the Poisson and the Nernst-Planck equations. Even for simple geometries, simultaneously solving these equations analytically is highly challenging.

In this section, the numerical procedure for the resolution of the electrokinetic governing equations mentioned above subject to appropriately assigned boundary conditions is introduced briefly, along with the calculation of the forces acting on a particle in electrophoresis. Here, the objective is not to assess the performance of various numerical approaches or how to select an efficient method; there exist many references in the literature for those purposes. Instead, we just illustrate, through some examples, the procedure for solving numerically typical electrophoresis problems.

Many sophisticated numerical techniques are available for the resolution of coupled partial differential equations. For illustration, the finite element method is chosen to solve the governing electrokinetic equations in our case. This, of course, does not imply that it is the best choice for the resolution of every electrophoresis problem. In principle, any numerical scheme that is capable of solving coupled partial differential equations, such as the finite element method, finite difference method, finite volume method, boundary element method, method of lines, and spectral method, to name a few, can be adopted [68]. However, because it is readily applicable to problems having complicated geometries and computational domains, the finite element method seems to be more efficient than other methods, especially for the case of a higher-dimensional problem. In the following discussion, the solution procedure is introduced for two general cases: problems involving high surface potential and those involving low surface potential.

6.6.2.1 High Surface Potential

For a more concise presentation, the general electrokinetic governing equations, Eqs. (6.39)–(6.44), are rewritten in scaled forms as

$$\nabla^{*2}\phi_1^* = -\frac{1}{(1+\gamma)}\frac{(\kappa a)^2}{\phi_r}[\exp(-\phi_r\phi_1^*) - \exp(\gamma\phi_r\phi_1^*)] \tag{6.187}$$

$$\nabla^{*2}\phi_2^* - \frac{(\kappa a)^2}{(1+\gamma)}[\exp(-\phi_r\phi_1^*) + \alpha\exp(\gamma\phi_r\phi_1^*)]\phi_2^*$$

$$= \frac{(\kappa a)^2}{(1+\gamma)}[\exp(-\phi_r\phi_1^*)g_1^* + \exp(\gamma\phi_r\phi_1^*)\gamma g_2^*] \tag{6.188}$$

$$\nabla^{*2}g_1^* - \phi_r\nabla^*\phi_1^* \cdot \nabla^*g_1^* = \phi_r^2 Pe_1 \mathbf{u}^* \cdot \nabla^*\phi_1^* \tag{6.189}$$

$$\nabla^{*2}g_2^* + \alpha\phi_r\nabla^*\phi_1^* \cdot \nabla^*g_2^* = \phi_r^2 Pe_2 \mathbf{u}^* \cdot \nabla^*\phi_1^* \tag{6.190}$$

$$\nabla^* \cdot \mathbf{u}^* = 0 \tag{6.191}$$

$$\nabla^{*2}\mathbf{u}^* - \nabla^*p^* + \nabla^{*2}\phi_1^*\nabla^*\phi_2^* + \nabla^{*2}\phi_2^*\nabla^*\phi_1^* = 0 \tag{6.192}$$

$$n_1^* = \exp(-\phi_r\phi_1^*)[1 - \phi_r(\phi_2^* + g_1^*)] \tag{6.193}$$

$$n_2^* = \exp(\gamma\phi_r\phi_1^*)[1 + \gamma\phi_r(\phi_2^* + g_2^*)] \tag{6.194}$$

In these expressions, $\nabla^* = a\nabla$, $\nabla^{*2} = a^2\nabla^2$, $\gamma = -z_2/z_1$, and $\phi_r = \zeta_k/(k_BT/ez_1)$, where $\zeta_k = \zeta_p$ if $\zeta_p \neq 0$, and $\zeta_k = \zeta_b$ if $\zeta_p = 0$; ζ_p and ζ_b are the surface potential of the particle and that of the boundary, respectively. $n_j^* = n_j/n_{j0}$, $\phi_j^* = \phi_j/\zeta_k$, and $g_j^* = g_j/\zeta_k$, $j = 1,2$. $\kappa = [\sum_{j=1}^2 n_{j0}(ez_j)^2/\varepsilon k_BT]^{1/2}$ is the reciprocal Debye length, $p^* = p/p_{ref}$, $p_{ref} = \varepsilon\zeta_k^2/a^2$, and $\mathbf{u}^* = \mathbf{u}/U_{ref}$, where $U_{ref} = (\varepsilon\zeta_k^2/\eta a)$ is the magnitude of the velocity of the particle predicted by Smoluchowski's theory when an electric field of strength (ζ_k/a) is applied. $Pe_j = \varepsilon(k_BT/z_1e)^2/\eta D_j$ is the electric Peclet number of ionic species j. As an example, for an aqueous solution of KCl at $T = 298$ K, $D_j \approx 2 \times 10^{-9}$ m²/s [84], $\varepsilon = 8.854 \times 10^{-12} \times 80$ C/(Vm), $k_BT/z_1e = 0.02568$ V, and $\eta = 0.993 \times 10^{-3}$ kg/(ms), yielding $Pe_j = 0.235$.

The procedure for solving Eqs. (6.187)–(6.194) subject to prespecified boundary conditions, such as Eqs. (6.50), (6.51), (6.53), (6.54), (6.66), and (6.71)–(6.77), or other types of conditions [56, 60] can be summarized as follows. Adopting the treatment of O'Brien and White [11], the present electrophoresis problem is first decomposed into two subproblems. In the first subproblem, a particle moves at a constant velocity in the absence of the applied electric field E_∞, and in the second subproblem, E_∞ is applied but the particle is fixed in the space. If we let F_i be the total force acting on the particle in subproblem i, and Fi be its magnitude, then $F_1 = \chi_1 U^*$ and $F_2 = \chi_2 E_\infty^*$, where the proportional constants χ_1 and χ_2 are independent of U^* and E_∞^*, respectively. Here, $E_\infty^* = E_\infty/(\zeta_k/a)$ and $U^* = U/U_{ref}$ are the scaled strength of the applied electric field and the scaled velocity, respectively. The fact that $F_1 + F_2 = 0$ at steady state yields [28]

$$\mu^* = \frac{U^*}{E_\infty^*} = -\frac{\chi_2}{\chi_1} \tag{6.195}$$

where μ^* is the scaled electrophoretic mobility of the particle. In the present case, the forces acting on the particle include the electrical force and the hydrodynamic force, and only the z components of these forces need to be considered. If we let $F_{E,i}$ and $F_{D,i}$ be the z components of the electrical force and the hydrodynamic force in subproblem i, respectively, then

$$F_i = F_{E,i} + F_{D,i}, \quad i = 1, 2 \tag{6.196}$$

As shown in the previous section, $F_{E,i}$ and $F_{D,i}$ can be evaluated by

$$F_{E,i}^* = \frac{F_{E,i}}{\varepsilon\zeta_k^2} = \iint\limits_{\Omega_p^*} (\sigma^{E*} \cdot \mathbf{n}) \cdot \mathbf{e}_z \, d\Omega_p^*$$

$$= \iint\limits_{\Omega_p^*} \left(\left[\frac{\partial\phi_1^*}{\partial n}\frac{\partial\phi_2^*}{\partial z^*} + \frac{\partial\phi_2^*}{\partial n}\frac{\partial\phi_1^*}{\partial z^*} \right] - \left[\frac{\partial\phi_1^*}{\partial n}\frac{\partial\phi_2^*}{\partial n} + \frac{\partial\phi_1^*}{\partial t}\frac{\partial\phi_2^*}{\partial t} \right] n_z \right) d\Omega_p^* \tag{6.197}$$

$$F_{D,i}^* = \frac{F_{D,i}}{\varepsilon\zeta_k^2} = \iint\limits_{\Omega_p^*} (\sigma^{H*} \cdot \mathbf{n}) \cdot \mathbf{e}_z \, d\Omega_p^* \tag{6.198}$$

where $F_{E,i}^*$ and $F_{D,i}^*$ are the scaled $F_{E,i}$ and $F_{D,i}$, respectively; Ω_p^* is the scaled particle surface; z^* is the scaled z component of the unit normal vector; and $\sigma^{E*} = \sigma^E/(\varepsilon\zeta_k^2/a^2)$ and $\sigma^{H*} = \sigma^H/(\varepsilon\zeta_k^2/a^2)$ are the scaled Maxwell stress tensor and the scaled shear stress tensor, respectively.

First, the governing equations for the electric and the flow fields are solved numerically subject to the associated boundary conditions, followed by the

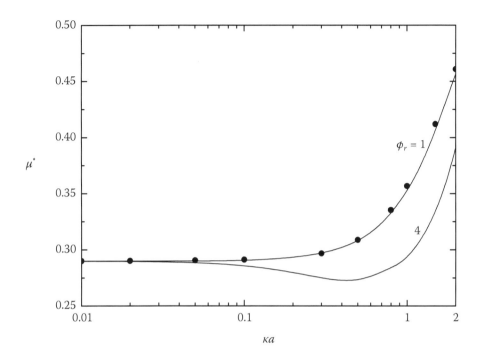

FIGURE 6.13

Variation of the scaled mobility μ^* as a function of κa at two levels of the scaled surface potential ϕ_r for the case of a positively charged sphere at the center of an uncharged spherical cavity at $\lambda = 0.4$ and $Pe_1 = Pe_2 = 0.1$.

Note: Solid curves: results based on FlexPDE; discrete symbols: analytical result of Zydney [25].

Source: Zydney, A. L., Boundary effects on the electrophoretic motion of a charged particle in a spherical cavity, *J. Colloid Interface Sci.*, 169, 476–485, 1995.

calculation of $F_{E,i}^*$ and $F_{D,i}^*$, $i = 1,2$. Then the scaled electrophoretic mobility of a particle is evaluated [11,28]. Several finite element method-based software programs are available, including ANSYS, Conventor, FEMLab, Fluent, FlexPDE, and ALGOR, to name a few. The implementation of a finite element method usually starts with the definition of a shape function and an element type. Typically, one discretizes a two-dimensional computational domain into triangular or quadrilateral elements. In FlexPDE, for instance, triangular elements are used.

Figure 6.13 shows an example of the results based on FlexPDE (version 2.22; PDE Solutions Inc., USA, 2000), where the scaled mobility μ^* is plotted against the thickness of the double layer κa at two levels of the scaled surface potential ϕ_r for the case of a positively charged sphere at the center of an uncharged spherical cavity. For comparison, the analytical results of Zydney [25], which are based on the Debye-Hückel condition, are also presented. As seen in Figure 6.13, the numerical results are consistent with the analytical result at a low surface potential ($\phi_r = 1$).

Figure 6.13 also reveals that the qualitative behavior of the mobility of a particle at a high surface potential ($\phi_r = 4$) is different from that at a low surface potential. In the latter, μ^* increases monotonically with increasing κa, but μ^* shows a local maximum as κa varies in the latter. A similar phenomenon was also observed by O'Brien and White [11] in a study of the electrophoresis of an isolated sphere in an infinite fluid medium. The difference in the electrophoretic behavior of a particle at high surface potentials and that at low surface potentials arises mainly from the effect of double-layer polarization (relaxation) [28,56]. In general, this effect becomes significant when the surface potential of a particle exceeds circa 75 mV ($\phi_r \cong 3$) and the thickness of the double layer is comparable to its linear size ($\kappa a \cong 1$).

6.6.2.2 Low Surface Potential

The scaled governing equations for the case of low surface potential can be summarized as below:

$$\nabla^{*2}\phi_1^* = (\kappa a)^2 \phi_1^* \tag{6.199}$$

$$\nabla^{*2}\phi_2^* = 0 \tag{6.200}$$

$$\nabla^* \cdot \mathbf{u}^* = 0 \tag{6.201}$$

$$0 = \nabla^{*2}\mathbf{u}^* - \nabla^* p^* + \nabla^{*2}\phi_1^* \nabla^* \phi_2^* \tag{6.202}$$

The procedure for solving Eqs. (6.199)–(6.202) subject to the boundary conditions stated previously; Eqs. (6.50) and (6.51) for the constant surface potential model; Eqs. (6.53) and (6.54) for the constant charge density model; Eq. (6.70) for the charge regulation model; Eqs. (6.71), (6.72), or (6.75) for perturbed electric potential; Eqs. (6.76) and (6.77) for the flow field; or other types of boundary conditions [57,61] can be summarized as follows. As in the case of high surface potential, we focused only on the z components of the electrostatic force F_E and the hydrodynamic force F_D acting on a particle. At steady state, the sum of these two forces must vanish, that is, $F_E + F_D = 0$. Based on this condition, a trial-and-error procedure [34–38] can be applied to estimate the electrophoretic velocity U. However, a more efficient approach is available where the cumbersome procedure can be avoided by mathematically decomposing the present problem into two subproblems [11,44]. In the first subproblem, the particle moves with U in the absence of the applied electric field $\mathbf{E}_\infty$, and in the second subproblem, $\mathbf{E}_\infty$ is applied, but the particle remains fixed. The first subproblem is of a purely hydrodynamic nature, where the particle experiences only a conventional hydrodynamic drag, $F_{D,1} = -UD$. Here, the drag coefficient D depends upon the geometry considered [79], and it is positive. In the second subproblem, because the particle is held fixed, it experiences an electrostatic force F_E coming from the applied $\mathbf{E}_\infty$. In addition, the flow of the mobile ions in the double layer surrounding the particle yields a hydrodynamic drag $F_{D,2}$ acting on the particle. At steady state, the sum of the above forces must vanish, that is, $F_E + F_D = 0$, where $F_D = F_{D,1} + F_{D,2}$, and therefore,

$$U = \frac{F_E + F_{D,2}}{D} \tag{6.203}$$

This relation suggests that the electrophoretic mobility of a particle can be evaluated by the following procedure.

1. Because $\phi_1^* = 0$ and $\phi_2^* = 0$ in the first subproblem, the electric body force $\nabla^{*2}\phi_1^*\nabla^*\phi_2^*$ can be removed from Eq. (6.202). In this case, assume an arbitrary value of U in Eq. (6.76) and solve Eqs. (6.201) and (6.202) simultaneously for the flow field.

2. Evaluate $F_{D,1}$ (or $D = F_{D,1}/U$) by Eq. (6.86).

3. In the second problem, ϕ_1^* and ϕ_2^* are evaluated first by solving Eqs. (6.199) and (6.200) subject to the assumed boundary conditions, such as Eqs. (6.50) and (6.51) for the case of the constant surface potential model, followed by evaluating F_E through Eq. (6.81).

4. Substitute the resultant ϕ_1^* and ϕ_2^* into Eq. (6.202), replace the boundary condition expressed in Eq. (6.76) by $\mathbf{u} = \mathbf{0}$, and solve Eqs. (6.201) and (6.202) simultaneously for the flow field.

5. Calculate $F_{D,2}$ by Eq. (6.86).

6. Substitute D, F_E, and $F_{D,2}$ into Eq. (6.203) to evaluate U.

Note that the solution procedures for the case of high surface potential and that of low surface potential are different. In the latter, the governing equations for the electric field and those used to evaluate $F_{D,1}$ (or D) are decoupled, which is not the case in the former. In addition, in the case of low surface potential, F_E, $F_{D,2}$, and $F_{D,1}$ (or D) need to be evaluated separately, which is different than in the case of high surface potential. In the case of low surface potential, the influence of each force (F_E, $F_{D,1}$, and $F_{D,2}$) on the electrophoretic behavior of a particle can be discussed, which becomes infeasible in the case of high surface potential.

Two examples are given to illustrate the numerical results based on FlexPDE: the electrophoresis of a positively charged sphere along the axis of an uncharged cylindrical pore, which is a symmetric system, and a positively charged sphere at an arbitrary position in an uncharged spherical cavity, which is an asymmetric system. For convenience, Eq. (6.203) is rewritten in scaled form as

$$U^* = \frac{F_E^* + F_{D,2}^*}{D^*} \tag{6.204}$$

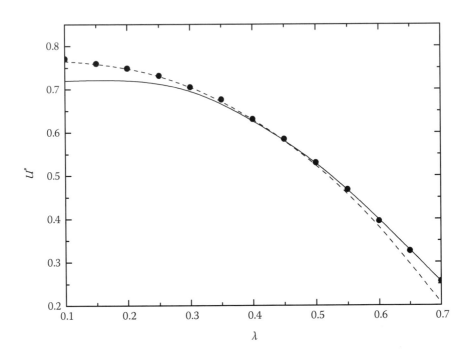

FIGURE 6.14

Variation of the scaled electrophoretic mobility U^* as a function of λ for the electrophoresis of a positively charged sphere along the axis of an uncharged cylindrical pore at $\kappa a = 4.3$.

Note: Solid curve: numerical result of Shugai and Carnie [29]; dashed curve: result of Ennis and Anderson [26] based on reflection method; discrete symbols: result based on FlexPDE.

Sources: Ennis, J., Anderson, J. L., Boundary effects on electrophoretic motion of spherical particles for thick double layers and low zeta potential, *J. Colloid Interface Sci.*, 185, 497–514, 1997; and Shugai, A. A., Carnie, S. L., Electrophoretic motion of a spherical particle with a thick double layer in bounded flows, *J. Colloid Interface Sci.*, 213, 298–315, 1999.

where $U^* = U/U_r$, $F_E^* = F_E/6\pi\eta a U_r$, $F_{D,2}^* = F_{D,2}/6\pi\eta a U_r$, and $D^* = D/6\pi\eta a$, with $U_r = \varepsilon\zeta_{ref}E/\eta$ and $\zeta_{ref} = k_B T/e$. Figure 6.14 shows the results based on FlexPDE, where the variation of the scaled electrophoretic mobility U^* is plotted against the degree of boundary effect, measured by the parameter λ (=radius of particle/radius of pore) [49]. The numerical results of Ennis and Anderson [26] and those of Shugai and Carnie [29] are also illustrated, for comparison. The former is based on a reflection method, and the latter based on the method of Teubner [85].

As pointed out by Shugai and Carnie [29], due to numerical error, their approach is inaccurate for smaller values of λ. On the other hand, the results of Ennis and Anderson [26] might be inaccurate for larger values of λ, where

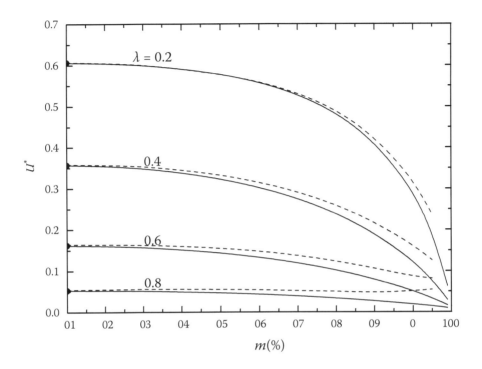

FIGURE 6.15

Variation of the scaled electrophoretic mobility U^* as a function of the position parameter m for the case of a positively charged sphere at an arbitrary position in an uncharged spherical cavity.

Note: The surface potential of the particle is 25.4 mV and $\kappa a = 1$. Solid curves: numerical result based on FlexPDE; dashed curve: Hsu et al. [35]; discrete symbol: analytical result of Zydney [25].

Source: Zydney, A. L., Boundary effects on the electrophoretic motion of a charged particle in a spherical cavity, *J. Colloid Interface Sci.*, 169, 476–485, 1995; and Hsu, J. P., Hung, S. H., Kao, C. Y., Electrophoresis of a sphere at an arbitrary position in a spherical cavity, *Langmuir*, 18, 8897–8901, 2002.

double-layer overlapping is serious. As seen in Figure 6.14, FlexPDE provides satisfactory results for the whole range of λ considered.

In the second example, the electrophoresis of a spherical particle in a spherical cavity is considered. Figure 6.15 shows the variation of the scaled electrophoretic mobility of the particle, U^*, as a function of the position parameter, m, at various values of λ (=particle radius/cavity radius). Here, $m(\%) = 100c/(b - a)\%$, with $a, b,$ and c being the radius of the particle, the radius of the cavity, and the center-to-center distance between the particle and the cavity, respectively. For comparison, the analytical results of Zydney [25] and the numerical results of Hsu et al. [35] are also presented in Figure 6.15. The former considered the case of $m = 0\%$ (sphere at the center of the cavity) only. The latter was based on a trial-and-error procedure where the way that the electrostatic force acting on a particle is evaluated is different from that in the present method. As seen in Figure 6.15, for $m = 0\%$ the present numerical results based on FlexPDE are consistent with the corresponding analytical results of Zydney [25], implying that the performance of the software adopted is satisfactory. Note that for smaller values of m, the results of Hsu et al. [35] are close to the present results. However, as m gets large, they are different both qualitatively and quantitatively, especially for larger values of λ. This is because, instead of Eqs. (6.48) and (6.81), Eqs. (6.49) and (6.84) are used in Hsu et al. [35], as in many other studies [7,27,38,40,67,68,75]. As pointed out by Hsu and Yeh [52], this yields an extra electric body force, $\rho_e \nabla \phi_1$, and an extra electrostatic force coming from the equilibrium electric potential, $\iint_S \varepsilon(\partial \phi_1/\partial n)(\partial \phi_1/\partial z)dS$, and, therefore, U^* is overestimated.

6.7 Concluding Remarks

Since the pioneer theoretical works of Helmholtz and Smoluchowski, electrophoresis, one of the most important electrokinetic phenomena, has been studied extensively, especially in the last few decades. Although deriving an analytical expression, which correlates the key parameters of a system, is highly desirable to, for example, experimentalists who are interested in knowing the surface properties of particles of colloidal size, engineers who are designing a separation-purification unit, and designers of an electrophoretic apparatus, it is still highly challenging under general conditions. Up to now, almost all of the available analytical results are based on drastic conditions, and therefore have limitations that may become unrealistic. It appears that the best strategy at the present stage is to choose an efficient and accurate numerical tool capable of predicting the electrophoretic behavior of a particle under conditions of practical significance. While this seems to be feasible on account of the capability of modern computing machines, unfortunately it can be nontrivial too. This is because factors such as the geometry of a problem; the shape, concentration, and physicochemical properties of particles; and the length scales involved make the resolution of the equations governing electrophoresis numerically difficult. For instance, available results for a three-dimensional problem are extremely limited in the literature. The contents presented in this chapter provide but a very limited picture of electrophoresis. Readers are recommended to check the literature constantly for newly published results, which seem to be quite a few in the last decade, for a more accurate expression which fits realistically to experimental conditions and/or for a more powerful analytical or numerical methodology.

List of Symbols

a linear size of a particle or radius of a sphere (m)

$A = e^2 z_1 N_{\Omega_p, A} a / \varepsilon k_B T$ (–)

A^- dissociated form of functional group HA (–)

$[A^-]$ number concentration of A^- per unit particle surface area ($1/m^2$)

A_1, A_2, A_3, A_4 integration constants (–)

AH functional group (–)

b distance between the center of a sphere to a boundary or radius of cavity (m)

B functional group (–)

B_1, B_2, B_3, B_4 integration constants (–)

$[B]$ numbers of B per unit particle surface area ($1/m^2$)

BH^+ associated form of functional group B (–)

$[BH^+]$ numbers of BH^+ per unit particle surface area ($1/m^2$)

$B = [H^+]_b / K_a$ (–)

c center-to-center distance between sphere and cavity (m)

D drag coefficient (kg/m)

$D^* = D/6\pi\eta a$, scaled drag coefficient (–)

D_j diffusivity of ionic species j (m^2/s)

e elementary charge (C)

$\mathbf{e}_z$ unit vector in z direction (–)

$\mathbf{e}_r$ unit vector in radial direction (–)

$f(ka)$ Henry's function defined by Eq. (6.4) (–)

$\mathbf{f}_j$ flux of ionic species j ($1/s/m^2$)

E_z strength of the local electric field in z direction (V/m)

E_∞ strength of the applied electric field (V/m)

$E_n(\kappa a)$ exponential integral of order n defined by Eq. (6.5) (–)

$E_\infty^* = E_\infty / (\zeta_k/a)$, scaled E_∞ (–)

$\mathbf{E}_\infty$ applied electric field (V/m)

F_D hydrodynamic force acting on a particle in z direction (N)

$F_{D,i}$ z component of hydrodynamic force in subproblem i defined by Eq. (6.198) (N)

$F_{D,i}^*$ scaled z component of hydrodynamic force in subproblem i defined by Eq. (6.198) (–)

F_E electric force acting on a particle in z direction (N)

$F_{E,i}$ z component of electric force in subproblem i defined by Eq. (6.197) (N)

$F_{E,i}^*$ scaled z component of electric force in subproblem i defined by Eq. (6.197) (–)

F_i electric force acting on a particle in z direction (N)

F_z total force acting on a particle defined by Eq. (6.183) (N)

g_j perturbed potential associated with ionic species j (V)

g_j^* scaled perturbed potential associated with ionic species j (–)

H^+ hydrogen ion (–)

$[H^+]_{\Omega_p}$ numbers of H^+ per unit particle surface area (1/m²)

$[H^+]_b$ bulk concentration of H^+ (1/m²)

$\mathbf{I}$ unit tensor (–)

j index for ionic species (–)

k_B Boltzmann constant (J/K)

K_a equilibrium dissociation constant defined by Eq. (6.57) (1/m²)

K_b equilibrium dissociation constant defined by Eq. (6.58) (1/m²)

$m = 100c/(b-a)\%$, position parameter (–)

n magnitude of unit normal vector $\mathbf{n}$ (–)

n_z z component of unit normal vector $\mathbf{n}$ (–)

n_j number concentration of ionic species j (1/m³)

n_{j0} bulk number concentration of ionic species j (1/m³)

n_j^* scaled number concentration of ionic species j (–)

$\mathbf{n}$ unit outward normal vector on particle surface (–)

N total number of species j (–)

$N_{\Omega_p,A}$ total number of the acidic functional group per unit area (1/m²)

$N_{\Omega_p,B}$ total number of the basic functional group per unit area (1/m²)

p perturbed hydrodynamic pressure of the liquid phase (Pa)

p_∞ equilibrium pressure of the fluid far from the particle (Pa)

$p_{ref} = \varepsilon \zeta_k^2 / a^2$, reference pressure (Pa)

p^* scaled hydrodynamic pressure (–)

P hydrodynamic pressure of the liquid phase (Pa)

Pe_j electric Peclet number of ionic species j (–)

q surface charge (C/m)

r radial coordinate (m)

t magnitude of unit tangential vector $\mathbf{t}$ (–)

t_z z component of unit tangential vector $\mathbf{t}$ (–)

t unit tangential vector on particle surface (–)

T absolute temperature (K)

T matrix transpose (–)

u_n normal component of liquid velocity on particle surface (m/s)

u_r liquid velocity in r direction (m/s)

u_t tangential component of liquid velocity on particle surface (m/s)

u_θ liquid velocity in θ direction (m/s)

u_∞ magnitude of liquid velocity far from the particle (m/s)

u velocity of liquid phase (m/s)

$\mathbf{u}_\infty$ liquid velocity far from the particle (m/s)

$\mathbf{u}^* = \mathbf{u}/U_{ref}$, scaled velocity of liquid phase (–)

U z component of particle velocity (m/s)

$U_r = \varepsilon(k_B T/e)E_\infty/\eta$ (m/s)

$U_{ref} = (\varepsilon\zeta_k^2/\eta a)(-)$

U^* scaled electrophoretic mobility defined by Eq. (6.204) (–)

$\mathbf{v}_j$ velocity of ionic species j (m/s)

z axial coordinate (m)

z^* scaled z component of unit normal vector (–)

z_j valence of ionic species j (–)

z_1 valence of cations (–)

z_2 valence of anions (–)

Z absolute value of the valence of ions in a $Z{:}Z$ electrolyte (–)

Greek Letters

α function defined in Eq. (6.184) (–)

β function defined in Eq. (6.185) (–)

$\gamma = -z_2/z_1$ (–)

∇ gradient operator (1/m)

∇^* scaled gradient operator (–)

∇^2 Laplace operator (1/m²)

∇^{*2} scaled Laplace operator (–)

ε permittivity of liquid phase (C²/N/m²)

$\zeta_k = \zeta_p$ if $\zeta_p \neq 0$, and $= \zeta_b$ if $\zeta_p = 0$ (V)

ζ_p surface potential of particle (V)

$\zeta_{ref} = k_B T/e$, reference potential (V)

ζ_w surface potential of boundary (V)

σ surface charge density (C/m)

σ_p charge density on particle surface defined in Eq. (6.12) (C/m)

$\sigma_{p,A}$ density of dissociated acidic functional groups (C/m)

$\sigma_{p,B}$ density of dissociation basic functional groups (C/m)

σ_b charge density on boundary surface (C/m)

σ^E Maxwell stress tensor (N/m²)

$\sigma^{E*} = \sigma^E/(\varepsilon\zeta_k^2/a^2)$, scaled Maxwell stress tensor (–)

σ^H hydrodynamic stress tensor (N/m²)

$\sigma^{H*} = \sigma^H/(\varepsilon\zeta_k^2/a^2)$, scaled hydrodynamic stress tensor (–)

μ electrophoretic mobility (m²/V/s)

μ^* scaled electrophoretic mobility defined in Eq. (6.195) (–)

η viscosity of liquid phase (kg /m/s)

κ reciprocal Debye length (m)

λ = (characteristic length of particle/characteristic length of boundary) (–)

θ angular coordinate (–)

ρ_e space charge density (C/m³)

ϕ electrical potential (V)

ϕ_1 equilibrium potential (V)

ϕ_1^* scaled equilibrium potential (–)

ϕ_2 perturbed potential (V)

ϕ_2^* scaled perturbed potential (–)

ϕ_r scaled surface potential of particle (–)

τ_{rr} normal shear stress on particle surface defined in Eq. (6.136) (N/m²)

$\tau_{r\theta}$ normal shear stress on particle surface defined in Eq. (6.137) (N/m²)

χ_1, χ_2 proportional constant

$\omega = \eta U/\varepsilon\zeta_p E_\infty$, scaled electrophoretic mobility (–)

$\Phi = K_a/K_b$ (–)

$\Omega = N_{\Omega_p,A}/N_{\Omega_p,B}$ (–)

O order of a function (–)

$X = \frac{\varepsilon E_\infty}{1-\lambda^3}$ (–)

φ function defined in Eq. (6.134) (–)

ψ function defined in Eq. (6.180) (–)

Ω_p surface of particle (m²)

Ω_p^* scaled particle surface (–)

Ω_b surface of boundary (m²)

Superscripts

(*e*) equilibrium quantity

* scaled quantity

Subscripts

(i) inner-layer property

(o) outer-layer property

Ω_p property on particle surface

Prefix

δ perturbed quantity

References

Chapters 1–5 References

1. Spasic, A. M., Mitrovic, M., Krstic, D. N., Classification of finely dispersed systems, and Spasic, A. M., Lazarevic, M. P., Krstic, D. N., Theory of electro-viscoelasticity, in: *Finely Dispersed Particles: Micro-, Nano, and Atto-Engineering*, A. M. Spasic and J. P. Hsu, Eds., CRC Press/Taylor & Francis, Boca Raton, FL, 2006, 1–23 and 371–393.

2. Malvern, E. L., *Introduction to the Mechanics of a Continuous Medium*, Prentice Hall, Englewood Cliffs, NJ, 1971, 1.

3. Stroeve, P., Frances, E., *Molecular Engineering of Ultra Thin Polymeric Films*, Elsevier, London, 1987.

4. Dresselhaus, M., Dresselhaus, G., Avouris, P., *Carbon Nanotubes, Synthesis: Structure, Properties and Applications*, Springer-Verlag, Berlin, 2001.

5. Dirac, P. A. M., *The Principles of Quantum Mechanics*, 4th ed., Clarendon, Oxford, 2003, chaps. 11–12.

6. Israelachvili, J. N., *Intermolecular and Surface Forces*, Academic Press, New York, 1992, 260–339.

7. Drexler, E. K., *Nanosystems: Molecular Machinery, Manufacturing and Computation*, Wiley, New York, 1992, 161–189.

8. Spasic, A. M., Electron transfer phenomenon at developed liquid/liquid interfaces: Fractional order time delay systems, 18th International Congress of Chemical and Process Engineering, Prague, Czech Republic, August 24–28, 2008, C7.1.

9. Viladsen, J., et al., Catalyst for dehydrogenating organic compounds, in particular, amines, thiols, and alcohols, and its preparation. U.S. Patent 4,224,190, September 23, 1980.

10. Mitrovic, M., Integrated dual and multiple separation systems applying combined separation techniques, in: *Precision Process Technology*, M. P. C. Weijnen and A. A. H. Drinkenburg, Eds., Dordrecht, Kluwer Academic Publisher, 1993, 393.

11. Spasic, A. M., Electroviscoelasticity of liquid-liquid interfaces, in: *Interfacial Electrokinetics and Electrophoresis*, A. V. Delgado, Ed., Marcel Dekker, New York, 2002, 837–867.

12. Spasic, A. M., Babic, M. D., Marinko, M. M., Djokovic, N. N., Mitrovic, M., Krstic, D. N., A new classification of finely dispersed systems, in: *Abstract of Papers*, part 5, 4th European Congress of Chemical Engineering, Granada, Spain, September 21–25, 2003, 5.2.39.

13. Lazarevic, M. P., Spasic, A. M., Electroviscoelasticity of liquid-liquid interfaces: Fractional approach, in: *Abstract of Papers*, part 5, 4th European Congress of Chemical Engineering, Granada, Spain, September 21–25, 2003, 5.2.33.

14. Spasic, A. M., Babic, M. D., Marinko, M. M., Djokovic, N. N., Mitrovic, M., Krstic, D. N., Classification of rigid and deformable interfaces in finely dispersed systems: micro-, nano-, and atto-engineering, in: *CD of Papers and Abstract of Papers*, part 2, 16th International Congress of Chemical and Process Engineering, Prague, Czech Republic, August 22–26, 2004, D5.1.

15. Spasic, A. M., Lazarevic, M. P., Theory of electroviscoelasticity: Fractional approach, in: *CD of Papers and Abstract of Papers*, part 2, 16th International Congress of Chemical and Process Engineering, Prague, Czech Republic, August 22–26, 2004, D5.4.

16. Probstein, R. F., *Physicochemical Hydrodynamics*, Wiley, New York, 1994, 352–369.

17. Kruyt, H. R., *Colloid Science*, Elsevier, Amsterdam, 1952, Vol. 1, 302–341.

18. Spasic, A. M., Krstic, D. N., Structure and stability of electrified liquid-liquid interfaces in fine dispersed systems, in: *Chemical and Biological Sensors and Analytical Electrochemical Methods*, A. J. Rico, M. A. Butler, P. Vanisek, G. Horval, A. E. Silva, Eds., ECS, Pennington, NJ, 1997, 415–426.

19. Reid, C. R., Prausnitz, J. M., Pauling, B. E., *The Properties of Gasses and Liquids*, McGraw-Hill, New York, 1989, 632–655.

20. Condon, E. U., and Odishaw, H., Eds., *Handbook of Physics*, McGraw-Hill, New York, 1958, sec. 4, part 13.

21. Pilling, M. J., *Reaction Kinetics*, Clarendon Press, Oxford, 1975, 37–48.

22. Yeremin, E. N., *The Foundation of Chemical Kinetics*, Mir Publishers, Moscow, 1979, 103–149.

23. Leonard, I. S., *Quantum Mechanics*, McGraw-Hill, New York, 1955, 7–90.

24. Krall, A. N., Trivelpiece, W. A., *Principles of Plasma Physics*, McGraw-Hill, New York, 1973, 98–128.

25. Gasser, R. P. H., Richards, W. G., *Entropy and Energy Levels*, Clarendon Press, Oxford, 1974, 27–38.

26. Hirchfelder, J. O., Curtiss, C. F., Bird, R. B., *Molecular Theory of Gases and Liquids*, Wiley, New York, 1954, 139.

27. Babic, M. D., et al., Characteristic parts of D2EHPA-TOPO process for uranium recovery from wet phosphoric acid, 9th International Congress of Chemical and Process Engineering, Prague, Czech Republic, August 30–September 4, 1987, D3.70.

28. Babic, M. D., Spasic, A. M., Marinko, M. M., Djokovic, N. N., *Phosphoric Fertilizers and Possibility of Its Elimination*, book 72, SANU, Belgrade, 1993, 41–55.

29. Pavasovic, V., Stevanovic, R., Prochazka, J., Influence of dispersed phase on power consumption in vibrating plate column, International Solvent Extraction Conference, Munich, Germany, 1986, Vol. 3, 107–113.

30. Holmes, J. H., Schafer, A. C., Some operating characteristics of the pump-mix mixer settler, *CEP*, 52, 201–204, 1956.

31. Siebenhofer, R., Marr, R., Modeling of mass transfer in solvent extraction, International Solvent Extraction Conference, Munich, Germany, 1986, Vol. 3, 281–284.

32. Moral, A., Cordero, G., Josa, J. M., The mixer-settler for uranium recovery from phosphoric acid by the D2EHPA-TOPO mixture, 2nd International Congress on Phosphoric Compounds, Boston, MA, April 21–25, 1979, 693–706.

33. Negoicic, D., Djokovic, N., Spasic, A., Babic, M., Tolic, A., Efficiency of some particular types of "pump-mix" mixer-settlers, in: *Proceedings of the 2nd Yugoslav Congress on Chemical Engineering and Process Technique*, Dubrovnik, Yugoslavia, 1987, 269.

34. Lawver, J. E., Elements of deceptive data presentation, *Mining Eng.*, 7, 1962, 46–51.

35. Spasic, A. M., Djokovic, N. N., Babic, M. D., Elements for designing of the entrained organic phase separation equipment, 4th Mediterranean Congress on Chemical Engineering, Barcelona, Spain, 1987, 372–373.

36. Spasic, A. M., Djokovic, N. N., Babic, M. D., Jovanovic, G. N., Population balance model in the breaking of emulsions, 11th International Congress of Chemical and Process Engineering, Prague, Czech Republic, August 29–September 3, 1993 D9.23.

37. Spasic, A. M., Djokovic, N. N., Babic, M. D., Jovanovic, G. N., Performance of demulsions: Predictions based on electromechanical principles, 11th International Congress of Chemical and Process Engineering, Prague, Czech Republic, August 29–September 3, 1993, D9.24.

38. Spasic, A. M., Jokanovic, V., Stability of the secondary droplet-film structure in polydispersed systems, *J. Colloid Interface Sci.*, 170, 229–240, 1995.

39. Spasic, A. M., Jokanovic, V., Krstic, D. N., Physical nature of the structure of electrified liquid-liquid interfaces, 187th ECS Meeting, Symposium on Liquid/Liquid Interfaces, Reno, Nevada, May 21–26, 1995, ECS/Interface 169.

40. Spasic, A. M., Electroviscoelasticity and double electrical layer, lecture at: Symposium on Selforganization of Nonequilibrium Processes, Ecka, DFS & DBS, 1995, 35–37.

41. Spasic, A. M., A theory of electroviscoelasticity: Structure and stability of fine dispersed systems, invited lecture at: Department of Physical Chemistry, University of Belgrade, Belgrade, 1995.

42. Babic, M. D., Spasic, A. M., Marinko, M. M., Djokovic, N. N., Pilot plant investigation of uranium extraction from phosphoric acid, in: *Geophysical Congress, Uranium-Mining and Hydrogeology*, B. Merkel, S., Hurst, E. P., Lohnert, W. Struckmeier, Eds., Bergakademie-Freiberg, Sachsen, Verlag Sven von Loga, Köln, 1995, 9–16.

43. Spasic, A. M., Djokovic, N. N., Canic, N., Babic, M. D., Tolic, A. S., Organic phase loop in the uranium recovery from wet phosphoric acid process, 9th International Congress of Chemical and Process Engineering, Prague, Czech Republic, August 30–September 4, 1987, D3.72.

44. Spasic, A. M., Djokovic, N. N., Canic, N., Babic, M. D., Decrease of entrainment losses in the uranium recovery from wet phosphoric acid process, International Solvent Extraction Conference, Moscow, Russia, July 18–24, 1988, Vol. 4, 227–230.

45. Spasic, A. M., Djokovic, N. N., Babic, M. D., Jovanovic, G. N., The secondary liquid-liquid nondestructive separation, *Chem. Biochem. Eng. Q.*, 5, 35–42, 1991.

46. Hancil, V., Rod, V., Reznickova, J., Measurement of coalescence in agitated dispersion by light transmittance technique, International Solvent Extraction Conference, Munich, Germany, 1986, Vol. 3, 81–88.

47. Spasic, A. M., Emulsions, in: A. M. Spasic, Ed., *Multiphase Dispersed Systems*, ITNMS, MNTRS, Belgrade, 1997, 1–46.
48. Spasic, A. M., Jokanovic, V., Krstic, D. N., A theory of electroviscoelasticity: A new approach for quantifying the behavior of liquid-liquid interfaces under applied fields, *J. Colloid Interface Sci.*, 186, 434–446, 1997.
49. Spasic, A. M., Krstic, D. N., Structure and stability of electrified liquid-liquid interfaces in fine dispersed systems, in: *Joint Meeting Proceedings of the International Society of Electrochemists and Electro-Chemical Society*, Paris, France, 1997, 415.
50. Spasic, A. M., A new topic in fine dispersed systems, 12th International Congress of Chemical and Process Engineering, Prague, Czech Republic, August 25–30, 1996, C1.6.
51. Spasic, A. M., Mechanism of the secondary liquid-liquid droplet-film rupture on inclined plate, *Chem. Eng. Sci.*, 47, 3949–3957, 1992.
52. Spasic, A. M., Krstic, D. N., A new approach to the existence of micro, nano, and atto dispersed systems, 13th International Congress of Chemical and Process Engineering, Prague, Czech Republic, August 23–28, 1988, CD 7.
53. Godfrey, J. C., Hanson, C., Slater, M. J., Tharmalingam, S., Studies of entrainment in mixer settlers, *AIChE Symp. Ser.*, 74, 127–133, 1978.
54. Godfrey, J. C., The formation of liquid-liquid dispersions-chemical and engineering aspects-flow phenomena of liquid-liquid dispersions in process equipment, *Inst. Chem. Eng.*, London, 1984, 1–10.
55. Tadros, T. F., Interfacial aspects of emulsification, *Inst. Chem. Eng.*, London, 1984, 11–28.
56. Hartland, S., Wood, S. M., The effect of applied force on drainage of the film between a liquid drop and horizontal surface, *AIChE J.*, 19, 810–817, 1983.
57. Spisak, W., Toroidal bubbles, *Nature*, 349, 23, 1991.
58. Spasic, A. M., Djokovic, N. N., Babic, M. D., Marinko, M. M., A new constitutive model of liquids, 13th International Congress of Chemical and Process Engineering, Prague, Czech Republic, August 23–28, 1998, CD7.
59. Oldshue, J. Y., *Fluid Mixing Technology*, McGraw-Hill, New York, 1983, 125–140.
60. Schramm, L. L., *Emulsions-Fundamentals and Application in Petroleum Industry*, American Chemical Society, Washington, DC, 1992, 131–170.
61. Mitrovic, M., Jaric, J., Nanocontinua, discontinua and spaces of interactions, lecture at Mathematical Faculty, University of Belgrade, Belgrade, 1997.
62. Spasic, A. M., Djokovic, N. N., Babic, M. D., Marinko, M. M., Jovanovic, G. N., Performance of demulsions: Entrainment problems in solvent extraction, *Chem. Eng. Sci.*, 52, 657–675, 1997.
63. Friberg, S. E., and Bothorel, R., Eds., *Microemulsions: Structure and Dynamics*, CRC Press, Boca Raton, FL, 1987, 173–195.
64. Prigogine, I., *The Molecular Theory of Solutions*, North Holland, New York, 1957.
65. Pao, Y. H., Hydrodynamic theory for the flow of viscoelastic fluid, *J. Appl. Phys.*, 28, 591–598, 1957.
66. Joseph, D. D., *Fluid Dynamics of Viscoelastic Liquids*, Springer-Verlag, New York, 1990, 539–596.
67. Grinfeld, M. A., Norris, A. N., Hamiltonian and onsageristic approaches in the nonlinear theory of fluid-permeable elastic continua, *Int. J. Eng. Sci.*, 35, 75–87, 1997.
68. Modern Plasma Physics, Trieste course 1979, IAEA, Vienna 1981, 249–273.

69. Prigogine, I., Defay, R., *Chemical Thermodynamics*, Longmans, London, 1965, 32–47, 437–449.
70. Devereaux, O. F., *Topics in Metallurgical Thermodynamics*, Wiley, New York, 1983, 333–372.
71. Karapetiantz, M., *Thermdynamique Chimique*, Mir Publishers, Moscow, 1978, 110–135.
72. Jost, W., *Diffusion*, Academic Press, New York, 1952, 436–479.
73. Rivet, A. C. D., *The Phase Rule and the Study of Heterogeneous Equilibrium*, Clarendon, Oxford, 1923, 25–29.
74. Perry, H. J., *Chemical Engineering Handbook*, McGraw-Hill, New York, 1941, 1213–1268.
75. Rosenow, W. M., Choi, H., *Heat, Mass and Momentum Transfer*, Prentice Hall, Englewood Cliffs, NJ, 1961, 24–29.
76. Schuhmann, R., Jr., *Metallurgical Engineering, Vol. 1, Engineering Principles*, Addison Wesley, Cambridge, MA, 1952, 143–146.
77. Turner, A. G., *Heat and Concentration Waves*, Academic Press, New York, 1972, 94–118.
78. Ono, S., Kondo, S., in: *Handbuch der Physik, Vol. 10, Structure of Liquids*, S. Flugge, Ed., Springer-Verlag, Berlin, 1960.
79. Hodgman, D. C., Weast, C. R., Selby, M. S., *Handbook of Chemistry and Physics*, 40th ed., Chem. Rubber, Cleveland, OH, 1958, 2512.
80. Davis, S. H., Contact line problems in fluid mechanics, *J. Appl. Mech.*, 50, 977–982, 1983.
81. Garstens, M. A., Noise in nonlinear oscillators, *J. Appl. Phys.*, 28, 352–356, 1957.
82. Orr, W. I., *Radio Handbook*, 17th ed., Editors and Engineers, New Augusta, IN, 1967, 60–65.
83. Ericksen, J. L., *Isledovanie po mehanike splosnih sred*, Mir Publishers, Moscow, 1977.
84. Andjelic, P. T., *Tenzorski racun*, N. K. Beograd, Belgrade, 1967, 263–266.
85. Jokanovic, V., Janackovic, D., Spasic, A. M., Uskokovic, D., Synthesis and formation mechanism of ultrafine spherical Al_2O_3 powders by ultrasonic spray pyrolisis, *Mater. Trans., JIM*, 37, 627–635, 1996.
86. Plavsic, M., *Mehanika viskoznih fluida*, PMF, University of Belgrade, 1986, 33–44.
87. Scherwood, J. D., Nitmann, J., Gradient governed growth: the effect of viscosity ratio on stochastic simulations of the Saffman-Taylor instability, *J. Physique*, 47, 15–22, 1986.
88. Torres, A., van Roij, R., Tellez, G., Finite thickness and charge relaxation in double-layer interactions, *J. Colloid Interface Sci.*, 301, 176–183, 2006.
89. Gallez, D., De Wit, A., Kaufman, M., Dynamics of a thin liquid film with a surface chemical reaction, *J. Colloid Interface Sci.*, 180, 524–536, 1996.
90. Kolar-Anic, L., Anic, S., Vukojevic, V., *Dinamika nelinearnih procesa: od monotone do oscilatorne evolucije*, Faculty of Physical Chemistry, University of Belgrade, 2004.
91. Nablle, J., Eldabe, T. M., Electrohydrodynamic stability of two superposed elasticoviscous liquids in plane Coutte flow, *J. Mat. Phys.*, 28, 2791–2800, 1987.
92. Takashima, M., Gosh, A. K., Electrohydrodynamic instability in a viscoelastic liquid layer, *J. Phys. Soc. Japan*, 47, 1717–1722, 1979.
93. Tabatabaei, S. M., van de Ven T. G. M., Rey, A. D., Electroviscous sphere-wall interactions, *J. Colloid Interface Sci.*, 301, 291–301, 2006.

94. Whitby, C. P., Djerdjev, A. M., Beattie, J. K., Warr, G. G., Nanoparticle adsorption and stabilization of surfactant-free emulsions, *J. Colloid Interface Sci.*, 301, 342–345, 2006.

95. Horanyi, G., Lang, G. G., Double-layer phenomena in electrochemistry: Controversial views on some fundamental notions related to electrified interfaces, *J. Colloid Interface Sci.*, 296, 1–8, 2006.

96. Chesters, A. K., The modelling of coalescence processes in fluid-liquid dispersions: A review of current understanding, *Trans. IChemE*, 69, part A, 259–270, 1991.

97. Spasic, A. M., Djokovic, N. N., Babic, M. D., Marinko, M. M., Krstic, D. N., The existence of fine dispersed systems based on a new constitutive model of liquids, 4th International Conference on Fundamental and Applied Aspects of Physical Chemistry, Belgrade, Yugoslavia, September 23–25, 1998, 556–558.

98. Ostojic, B., Peric, M., Radic-Peric, *Ab initio* treatment of the Renner-Tellerovag effect in tetra-atomic molecules, *Centennial of the Serbian Chem. Soc., J. Serb. Chem. Soc.*, September 23–27, 1997, 77.

99. Spasic, A. M., Marinko, M. M., Babic, M. D., Djokovic, N. N., Jovanic, P., Jokanovic, V., Vunjak, N. V., Jovanovic, G. N., Krstic, D. N., Fine dispersed systems: Emulsions, *Centennial of the Serbian Chem. Soc., J. Serb. Chem. Soc.*, September 23–27, 1997, 78.

100. Misek, T., Components of the coalescence process in dense dispersions, International Solvent Extraction Conference, Munich, Germany, 1986, Vol. 3, 71–79.

101. Christensen, R. M., *Theory of Viscoelasticity: An Introduction*, Academic Press, New York, 1971.

102. Spasic, A. M., Krstic, D. N., Theory of electroviscoelasticity: A new approach to the existence of charged interfaces in micro, nano, and atto dispersed systems, lecture at Faculty of Mathematics, University of Belgrade, Belgrade, Yugoslavia, 1997.

103. Kolar-Anic, Lj., *Osnove Statisticke Termodinamike*, Faculty of Physical Chemistry, University of Belgrade, Belgrade, 1995, 174–184.

104. Millikan, R. A., *Electrons (+ and –), Protons, Neutrons, Mesotrons, and Cosmic Rays*, 1934, Prosveta, Belgrade, trans. 1948, 249–275.

105. Osipov, I. L., *Surface Chemistry*, Reinhold, New York, 1964, 295–340.

106. Adamson, A., *Physical Chemistry of Surfaces*, Wiley, New York, 1967, 505–533.

107. Lo, T. C., Baird, M., Hanson, C., Eds., *Handbook of Solvent Extraction*, Wiley, New York, 1983, 275–514.

108. Jaric, J., *Mehanika kontinuuma*, G. K. Beograd, Belgrade, 1988, 117–159.

109. Pugh, R. J., Foaming, foam films, antifoaming and defoaming, *Adv. Colloid Interface Sci.*, 64, 67–142, 1996.

110. Oldham, K. B., Spanier, J., *The Fractional Calculus*, Mathematics in Science and Engineering series, Vol. 111, Academic Press, New York, 1974.

111. Podlubny, I., *Fractional Differential Equations*, Academic Press: San Diego, CA, 1999.

112. Mainardi, F., Fractional calculus: Some basic problems in continuum and statistical mechanics, in: *Fractals and Fractional Calculus in Continuum Mechanics*, A. Carpinteri and F. Mainardi, Eds., Springer, Vienna, 1997, 291–348.

113. Sakakibara, S. Properties of vibration with fractional derivative damping of order 1/2, *JSME Int. J.*, C40, 393–399, 1997.

114. Bagley, R. L., Calico, R. A., Fractional order state equations for the control of viscoelasticity damped structures, *J. Guidance*, 14, 1412–1417, 1991.
115. Oustaloup, A., *La Commande CRONE*, Hermes, Paris, 1991.
116. Kilbas, A., Srivastava, H., Truhillo, J., *Theory and Applications of Fractional Differential Equations*, Elsevier, Amsterdam, 2006.
117. Lazarević, P. M., D^α type iterative learning control for fractional LTI system, in: *Proc. ICCC2003*, Tatranska Lomnica, Slovak Republic, May 26–29, 2003, 869.
118. Caputo, M., *Elasticita e Dissipazione*, Zanichelli, Bologna, Italy, 1969.
119. Babenko, Yu., *Heat and Mass Transfer*, Chimia, Leningrad, 1986.
120. Torvik, P. J., Bagley, R. L., A theoretical basis for the application of fractional calculus to viscoelasticity, *J. Rheol.*, 27, 201–210, 1983.
121. Giannantoni, C., The problem of the initial conditions and their physical meaning in linear differential equations of fractional order, *Appl. Math. Comput.*, 141, 87–102, 1–15, 2003.
122. Diethelm, K., Ford, N. J., Analysis of fractional differential equations, *J. Math. Anal.*, 265, 401–418, 2002.
123. El-Sayed, A., Multivalued fractional differential equation, *J. Appl. Math. Comput.*, 68, 15–25, 1995.
124. Lubich, C., Discretized fractional calculus, *SIAM J. Math Anal.*, 17, 704–719, 1985.
125. Lubich C., A stability analysis of convolution quadratures for Abel-Volterra integral equations of the second kind, *IMA J. Numer. Anal.*, 6, 463–469, 1986.
126. Schmidt, V. H., Drumheller, J. E., Dielectric properties of lithium hydrazinium sulfate, *Phys. Rev. B*, 4, 4582–4597, 1971.
127. Tarasov, E., Zaslavsky, G., Dynamics with low-level fractionality, *Physica A*, 368, 399–415, 2006.
128. Mark, R. J., Hall, M. W., Differintegral interpolation from a bandlimited signal's samples, *IEEE Trans. Acoust. Speech Signal Process.*, 29, 872–877, 1981.
129. Bagley, R. L., Torvik, P. J., A theoretical basis for the application of fractional calculus to viscoelasticity, *J. Rheol.*, 27, 201–210, 1983.
130. Caputo, M., Linear models of dissipation whose Q is almost frequency inde-pent-2, *Geophys. J. Roy. Astron. Soc.*, 13, 529–539, 1967.
131. Ichise, M., Nagayanagi, Y., Kojima, T., Analog simulation of non-integer order transfer functions for analysis of electrode processes, *J. Electroanal. Chem. Interfacial Electrochem.*, 33, 253–265, 1971.
132. Oldham, K. B., A signal-independent electroanalytical, *Anal. Chem.*, 44, 196–198, 1972.
133. Hilfer, R., *Applications of Fractional Calculus in Physics*, World Scientific, Singapore, 2000.
134. Diethelm, K., Ford, N. J., Freed, A. D., Luchko, Y., Algorithms for the fractional calculus: A selection of numerical methods, *Comput. Methods Appl. Mech. Eng.*, 194, 743–773, 2005.
135. Heymans, N., Podlubny, I., Physical interpretation of initial conditions for fractional differential equations with Riemann-Liouville fractional derivatives, *Rheol. Acta*, 45, 765–771, 2006.
136. Spasic, A. M., Lazarevic, M. P., Electroviscoelasticity of liquid/liquid interfaces: Fractional-order model, *J Colloid Interface Sci.*, 282, 223–230, 2005.

137. Spasic, A. M., Lazarevic, M. P., Electron transfer at developed interfaces: Electroviscoelastic phenomena, 7th World Congress of Chemical Engineering, Glasgow, Scotland, July 10–14, 2005, 506.

138. Lazarevic, M. P., Spasic, A. M., Finite time stability analysis of linear nonautonomous fractional order systems with delayed states: Bellman-Gronwall approach, 7th World Congress of Chemical Engineering, Glasgow, Scotland, July 10–14, 2005, 288.

139. Spasic, A. M., Lazarevic, M. P., Mitrovic, M. V., Krstic, D. N., Electron and momentum transfer phenomena at developed deformable and rigid liquid-liquid interfaces, 1st South East-European Congress of Chemical Engineering, Belgrade, Serbia and Montenegro, September 25–28, 2005, 125.

140. Lazarevic, M. P., Spasic, A. M., Fractional order model of electroviscoelastic liquid-liquid interfaces: Nonlinear case, 25th Yugoslav Congress of Theoretical and Applied Mechanics, Novi Sad, Serbia and Montenegro, June 1–3, 2005, 86.

141. Lazarevic, M. P., Spasic, A. M., Fractional calculus applied to the theory of electroviscoelasticity, in: *Contemporary Problems in Civil Engineering*, Subotica, Serbia, June 2–3, 2006.

142. Spasic, A. M., Lazarevic, M. P., Mitrovic, M. V., Krstic, D. N., Electron and momentum transfer phenomena at developed deformable and rigid liquid-liquid interfaces, *CI&CEQ*, 12, 123–132, 2006.

143. Spasic, A. M., Phylosophical breakpoints related to micro-, nano-, and atto-engineering, 16th International Congress of Chemical and Process Engineering, Prague, Czech Republic, August 27–31, 2006, C7.1.

144. Lazarevic, M. P., Spasic, A. M., Electroviscoelasticity of liquid-liquid interfaces: Fractional order van der Pol model, linear and nonlinear case, in 16th International Congress of Chemical and Process Engineering, Prague, Czech Republic, August 27–31, 2006, C 7.5.

145. Spasic, A. M, Classification of finely dispersed systems: Philosophical breakpoints, 8th International Conference on Fundamental and Applied Aspects of Physical Chemistry, Belgrade, Serbia, September 26–29, 2006, 486–488.

146. Jokanovic, V., Dramicanin, M. D., Andric, Z., Jokanovic, B., Nedic, Z., Spasic, A. M., Luminescence properties of SiO_2:Eu^{3+} nanopowders: Multi-step nanodesigning, *J. Alloys Compounds*, 453, 253–260, 2008.

147. Saboni, A., Alexandrova, S., Spasic, A. M., Gourdon, C., Effect of the viscosity ratio on mass transfer from a fluid sphere at low to very high Peclet numbers, *Chem. Eng. Sci.*, 62, 4742–4750, 2007.

Chapter 6 References

1. Reuss, F. F., *Mém. Soc. Impériale. Naturalistes de Moscou*, 2, 327–344, 1809.

2. Quincke, G., Über die Fortfüjung materielle Theilchen durch strömende Elektrictät, *Pogg. Ann.*, 113, 513–598, 1861.

3. von Helmholtz, H, Studies of electric boundary layers, *Ann. Phys. Chem.*, 7, 337–387, 1879.

4. von Smoluchowski, M., Versuch einer mathematischen Theorie der Koagulationskinetik kolloider lösungen, *Z. Phys. Chem.*, 92, 129–168, 1918.
5. Hückel, E. Die kataphorese der kugel, *Physik. Z.*, 25, 204–210, 1924.
6. Henry, D. C., The cataphoresis of suspended particles, Part 1. The equation of cataphoresis, *Proc. Roy. Soc. London Ser. A*, 133, 106–129, 1931.
7. Masliyah, J. H., *Electrokinetic Transport Phenomena*, Alberta Oil Sands Technology and Research Authority, Edmonton, Canada, 1994.
8. Overbeek, J. T. G., Theory of the relaxation effect in electrophoresis, *Kolloide Beihefte*, 54, 287–364, 1943.
9. Booth, F., The cataphoresis of spherical, solid non-conducting particles in a symmetrical electrolyte, *Proc. Roy. Soc. London Ser. A*, 203, 514, 1950.
10. Wiersema, P. H., Loeb, A. L., Overbeek, J. T. G., Calculation of the electrophoretic mobility of a spherical colloid particle, *J. Colloid Interface Sci.*, 22, 78–99, 1966.
11. O'Brien, R. W., White, L. R., Electrophoretic mobility of a spherical colloidal particle, *J. Chem. Soc. Faraday Trans. 2*, 74, 1607–1626, 1978.
12. Nordsieck, A., Numierical integration of ordinary difference equations, *Math. Comput.*, 16, 22–49, 1962.
13. Morrison, F. A., Stukel, J. J., Electrophoresis of an insulating sphere normal to a plane, *J. Colloid Interface Sci.*, 33, 88–93, 1970.
14. Keh, H. J., Anderson, J. L., Boundary effects on electrophoretic motion of colloidal spheres, *J. Fluid Mech.*, 153, 417–439, 1985.
15. Keh, H. J., Chen, S. B., Electrophoresis of a colloidal sphere parallel to a dielectric plane, *J. Fluid Mech.*, 194, 377–390, 1988.
16. Keh, H. J., Lien, L. C., Electrophoresis of a dielectric sphere normal to a large conducting plane, *J. Chin. Inst. Chem. Eng.*, 20, 283–290, 1989.
17. Keh, H. J., Lien, L. C., Electrophoresis of a colloidal sphere along the axis of a circular orifice or a circular disk, *J. Fluid Mech.*, 224, 305–333, 1991.
18. Keh, H. J., Horng, K. D., Kuo, J., Boundary effect on electrophoresis of colloidal cylinders, *J. Fluid Mech.*, 231, 211–228, 1991.
19. Feng, J. J., Wu, W. Y., Electrophoretic motion of an arbitrary prolate body of revolution toward an infinite conducting wall, *J. Fluid Mech.*, 264, 41–58, 1994.
20. Chen, S. B., Koch, D. L., Electrophoresis and sedimentation of charged fibers, *J. Colloid Interface Sci.*, 180, 466–477, 1996.
21. Keh, H. J., Jan, J. S., Boundary effect on diffusiophoresis and electrophoresis: Motion of a colloidal sphere normal to a plane wall, *J. Colloid Interface Sci.*, 183, 458–475, 1996.
22. Keh, H. J., Chiou, J. Y., Electrophoresis of a colloidal sphere in a circular pore, *AIChE J.*, 224, 305–333, 1996.
23. Yariv, E., Brenner, H., The electrophoretic mobility of an eccentrically positioned spherical particle in a cylindrical pore, *Phys. Fluids*, 14, 3354–3357, 2002.
24. Ye, C., Sinton, D., Erickson, D., Li, D., Electrophoretic motion of a circular cylindrical particle in a circular cylindrical microchannel, *Langmuir*, 18, 9095–9101, 2002.
25. Zydney, A. L., Boundary effects on the electrophoretic motion of a charged particle in a spherical cavity, *J. Colloid Interface Sci.*, 169, 476–485, 1995.
26. Ennis, J., Anderson, J. L., Boundary effects on electrophoretic motion of spherical particles for thick double layers and low zeta potential, *J. Colloid Interface Sci.*, 185, 497–514, 1997.

27. Lee, E., Chu, J. W., Hsu, J. P., Electrophoretic mobility of a spherical particle in a spherical cavity, *J. Colloid Interface Sci.*, 196, 316–320, 1997.

28. Lee, E., Chu, J. W., Hsu, J. P., Electrophoretic mobility of a sphere in a spherical cavity, *J. Colloid Interface Sci.*, 205, 65–76, 1998.

29. Shugai, A. A., Carnie, S. L., Electrophoretic motion of a spherical particle with a thick double layer in bounded flows, *J. Colloid Interface Sci.*, 213, 298–315, 1999.

30. Chu, J. W., Lin, W. H., Lee, E., Hsu, J. P., Electrophoresis of a sphere in a spherical cavity at arbitrary electrical potentials, *Langmuir*, 20, 6289–6297, 2001.

31. Tang, Y. P., Chih, M. H., Lee, E., Hsu, J. P., Electrophoretic motion of a charge-regulated sphere normal to a plane, *J. Colloid Interface Sci.*, 242, 121–126, 2001.

32. Chin, M. H., Lee, E., Hsu, J. P., Electrophoresis of a sphere normal to a plane at arbitrary electrical potential and double layer thickness, *J. Colloid Interface Sci.*, 248, 383–388, 2002.

33. Lee, E., Kao, J. D., Hsu, J. P., Electrophoresis of a nonrigid entity in a spherical cavity, *J. Phys. Chem. B*, 106, 8790–8795, 2002.

34. Hsu, J. P., Kao, C. Y., Electrophoresis of a finite cylinder along the axis of a cylindrical pore, *J. Phys. Chem. B*, 106, 10605–10609, 2002.

35. Hsu, J. P., Hung, S. H., Kao, C. Y., Electrophoresis of a sphere at an arbitrary position in a spherical cavity, *Langmuir*, 18, 8897–8901, 2002.

36. Hsu, J. P., Hung, S. H., Electrophoresis of a charge-regulated spheroid along the axis of a cylindrical pore, *J. Colloid Interface Sci.*, 264, 121–127, 2003.

37. Hsu, J. P., Hung, S. H., Electrophoresis of a spheroid in a spherical cavity, *Langmuir*, 19, 7469–7473, 2003.

38. Hsu, J. P., Hung, S. H., Kao, C. Y., Tseng, S., Electrophoresis of a spheroid along the axis of a cylindrical pore, *Chem. Eng. Sci.*, 58, 5339–5347, 2003.

39. Liu, H., Bau, H., Hu, H., Electrophoresis of concentrically and eccentrically positioned cylindrical particles in a long tube, *Langmuir*, 20, 2628–2639, 2004.

40. Hsu, J. P., Ku, M. H., Kao, C. Y., Electrophoresis of a spherical particle along the axis of a cylindrical pore: Effect of osmotic flow, *J. Colloid Interface Sci.*, 276, 248–254, 2004.

41. Lee, E., Tang, Y. P., Hsu, J. P., Electrophoresis of a membrane-coated sphere in a spherical cavity, *Langmuir*, 20, 9415–9421, 2004.

42. Yu, H. Y., Hung, S. H., Hsu, J. P., Electrophoresis of a charge-regulated particle at an arbitrary position in a spherical cavity, *Colloid Polym. Sci.*, 283, 10–14, 2004.

43. Ye, C., Li, D., Electrophoretic motion of two spherical particles in a rectangular microchannel, *Microfluid Nanofluid*, 1, 52–61, 2004.

44. Hsu, J. P., Ku, M. H., Boundary effect on electrophoresis: Finite cylinder in a cylindrical pore, *J. Colloid Interface Sci.*, 283, 592–600, 2005.

45. Ye, C., Xuan, X., Li, D., Eccentric electrophoretic motion of a sphere in circular cylindrical microchannels, *Microfluid Nanofluid*, 1, 234–241, 2005.

46. Chen, P. Y., Keh, H. J., Diffusiophoresis and electrophoresis of a charged sphere parallel to one or two plane walls, *J. Colloid Interface Sci.*, 286, 774–791, 2005.

47. Hsu, J. P., Ku, M. H., Kuo, C. C., Electrophoresis of a charge-regulated sphere normal to a large disk, *Langmuir*, 21, 7588–7597, 2005.

48. Hsu, J. P., Kuo, C. C., Ku, M. H., Electrophoresis of a toroid along the axis of a cylindrical pore, *Electrophoresis*, 27, 3155–3165, 2006.

49. Hsu, J. P., Kuo, C. C., Electrophoresis of a finite cylinder positioned eccentrically along the axis of a cylindrical pore, *J. Phys. Chem. B*, 110, 17607–17615, 2006.

50. Davison, S. M., Sharp, K. V., Boundary effects on the electrophoretic motion of cylindrical particles: Concentrically-positioned particles in a capillary, *J. Colloid Interface Sci.*, 303, 288–297, 2006.
51. Qian, S. Z., Wang, A. H., Afonien, A. K., Electrophoretic motion of a spherical particle in a converging-diverging nanotube, *J. Colloid Interface Sci.*, 303, 579–592, 2006.
52. Hsu, J. P., Yeh, L. H., Comparison of three methods for the evaluation of the electric force on a particle in electrophoresis, *J. Chin. Inst. Chem. Eng.*, 37, 601–607, 2006.
53. Hsu, J. P., Yeh, L. H., Chen, Z. S., Effect of a charged boundary on electrophoresis: A sphere at arbitrary position in a spherical cavity, *J. Colloid Interface Sci.*, 310, 281–291, 2007.
54. Hsu, J. P., Yeh, L. H., Electrophoresis of two identical rigid spheres in a charged cylindrical pore, *J. Phys. Chem. B*, 111, 2579–2586, 2007.
55. Hsu, J. P., Chen, Z. S., Electrophoresis of a sphere along the axis of a cylindrical pore: Effect of double-layer polarization and electroosmotic flow, *Langmuir*, 23, 6198–6204, 2007.
56. Hsu, J. P., Chen, Z. S., Ku, M. H., Yeh. L. H., Effect of a charged boundary on electrophoresis: Sphere in spherical cavity at arbitrary potential and double-layer thickness, *J. Colloid Interface Sci.*, 314, 256–263, 2007.
57. Hsieh, T. H., Keh, H. J., Boundary effect on electrophoresis of a colloidal cylinder with a nonuniform zeta potential distribution, *J. Colloid Interface Sci.*, 315, 343–354, 2007.
58. Keh, H. J., Hsieh, T. H., Electrophoresis of a colloidal sphere in a spherical cavity with arbitrary zeta potential distributions and arbitrary double-layer thickness, *Langmuir*, 24, 390–398, 2008.
59. Hsu, J. P., Kuo, C. C., Ku, M. H., Electrophoresis of a charge-regulated toroid normal to a large disk, *Electrophoresis*, 29, 348–357, 2008.
60. Tseng, S., Cho, C. H., Chen, Z. S., Hsu, J. P., Electrophoresis of an ellipsoid along the axis of a cylindrical pore: Effect of a charged boundary, *Langmuir*, 24, 2929–2937, 2008.
61. Qian, S. Z., Joo, S. W., Hou, W. S., Zhao, X. X., Electrophoretic motion of a spherical particle with a symmetric, nonuniform surface charge distribution in a nanotube, *Langmuir*, 24, 5332–5340, 2008.
62. Hsu, J. P., Chen, C. Y., Lee, D. J., Tseng. S., Su, A., Electrophoresis of a charge-regulated sphere at an arbitrary position in a charged spherical cavity, *J. Colloid Interface Sci.*, 325, 516–525, 2008.
63. Hsu, J. P., Chen, Z. S., Effect of double-layer polarization and electroosmotic flow on the electrophoresis of an ellipsoid in a spherical cavity, *J. Phys. Chem. B*, 112, 11270–11277, 2008.
64. Hsu, J. P., Chen, Z. S., Lee, D. J., Tseng. S., Su, A., Effect of double-layer polarization and electroosmotic flow on the electrophoresis of finite cylinder along the axis of a cylindrical pore, *Chem. Eng. Sci.*, 63, 4561–4569, 2008.
65. Wu, Z., Gao, Y., Li, D., Electrophoretic motion of ideally polarizable particles in a microchannel, *Electrophoresis*, 30, 773–781, 2009.
66. Hsu, J. P., Yeh, L. H., Ku, M. H., Evaluation of the electric force in electrophoresis, *J. Colloid Interface Sci.*, 305, 324–329, 2007.
67. Ohshima, H., Electrophoresis of soft particles, *Adv. Colloid Interface Sci.*, 62, 189–325, 1995.

68. Masliyah, J. H., Bhattacharjee, S., *Electrokinetic Transport Phenomena*, 4th ed, Wiley, New York, 2006.

69. Hunter, R. J., *Foundations of Colloid Science*, Vol. 1, Oxford University Press, Oxford, 1989.

70. Bello, M. S., de Besi, P., Rezzonico, R., Righetti, P. G., Gasiraghi, E., Electroosmosis of polymer solutions in fused-silica capillaries, *Electrophoresis*, 15, 623–626, 1994.

71. Corradini, D., Buffer additives other than the surfacant sodium dodecyl sulfate for protein separations by capillary electrophoresis, *J. Chromatogr. B*, 699, 221–256, 1997.

72. Belder, D., Ludwig, M., Surface modification in microchip electrophoresis, *Electrophoresis*, 24, 3595–3606, 2003.

73. Hsu, J. P., Yeh, L. H., Electrophoresis of a charged boundary on electrophoresis in a Carreau fluid: A sphere at an arbitrary position in a spherical cavity, *Langmuir*, 23, 8637–8646, 2007.

74. Ninham, B. W., Parsegian, V. A., Electrostatic potential between surfaces bearing ionizable groups in ionic equilibrium with physiologic saline solution, *J. Theor. Biol.*, 31, 405–428, 1971.

75. Hunter, R. J., *Foundations of Colloid Science*, 2nd ed., Oxford University Press, Oxford, 2001.

76. Levine, S., Neale, G. H., The prediction of electrokinetic phenomena with multiparticle systems. I. Electrophoresis and electroosmosis, *J. Colloid Interface Sci.*, 47, 520–529, 1974.

77. Hsu, J. P., Yeh, L. H., Ku, M. H., Electrophoresis of a spherical particle along the axis of a cylindrical pore filled with a Carreau fluid, *Colloid Polym. Sci.*, 284, 886–892, 2006.

78. Hsu, J. P., Yeh, L. H., Yeh, S. J., Electrophoresis of a rigid sphere in a Carreau fluid normal to a large charged disk, *J. Phys. Chem. B*, 111, 12351–12361, 2007.

79. Happel, J., Brenner, H., *Low Reynolds Number Hydrodynamics*, Academic Press, New York, 1983.

80. Backstrom, G., *Fluid Dynamics by Finite Element Analysis*, Studentlitteratur, Sweden, 1999.

81. Shaw, D. J., *Introduction to Colloid and Surface Chemistry*, 3rd ed., Butterworths, London, 1980.

82. O'Brien, R. W., Hunter, R. J., The electrophoretic mobility of a large colloidal particles, *Can. J. Chem.*, 59, 1878–1887, 1981.

83. O'Brien, R. W., The solution of the electrokinetic equations for colloidal particles with thin double layers, *J. Colloid Interface Sci.*, 92, 204–216, 1983.

84. Chun, B., Ladd, A. J. C., The electroviscous force between charged particles: Beyond the thin-double-layer approximation, *J. Colloid Interface Sci.*, 274, 687–694, 2004.

85. Teubner, M., The motion of charged colloidal particles in electric fields, *J. Chem. Phys.*, 76, 5564–5573, 1982.

Author Index

Subject Index